Gotha Aircraft of WWI

A Centennial Perspective on Great War Airplanes

Jack Herris

Great War Aviation Centennial Series #6

This book is dedicated to my friend and colleague Colin Owers, whose generous help with photographs, drawings, and feedback has been indispensible to this series of books.

ACKNOWLEDGEMENTS
My sincere thanks to Colin Owers for photographs and his scale drawings of the LE3, LE4, UWD, and WD7, Daniel Brackx for permission to use his Gotha G.IX(LVG) photographs, Greg VanWyngarden for his photographs, Bob Pearson for his color profiles, Aaron Weaver for photographs and image editing, Martin Digmayer for his WD11 and WD14 drawings, and the Deutsches Technikmuseum in Berlin for photographs. Thanks also to Ron Kintner for his very helpful comments on Geman naval aviation unit designations and the use of Gotha seaplanes. Finally, Colin Owers acknowledges the researches of RM Zasadney for the Gotha *Taube* drawings. Any errors are my responsibility.

Composite cover photograph created and edited by Aaron Weaver.

Color aircraft profiles © Bob Pearson. Purchase his CD of WWI aircraft profiles for $50 US/Canadian, 40 €, or £30, airmail postage included, via Paypal to Bob at: **bpearson@kaien.net**

For our aviation books in print and electronic format, please see our website at: **www.aeronautbooks.com**.
I am looking for photographs of the less well-known German aircraft of WWI. For questions or to help with photographs you may contact me at **jherris@verizon.net**.

ISBN: **978-1-935881-14-8**

Design and layout: Jack Herris
Cover design: Jack Herris
Digital photo editing: Jack Herris & Aaron Weaver

Aeronaut Books

www.aeronautbooks.com

Table of Contents

Gotha WD11 in flight with naval identification streamers attached. Designed and successfully operated as a torpedo bomber, the WD11 was also used for long-range reconnaissance and bombing.

Gothaer Waggonfabrik AG – Gotha

Above: A Gotha G.IV of *Kagohl* 3, the *England Geschwader*, with its crew. Gotha was a relatively obscure manufacturer until the dramatic daylight bombing raid on London by Gotha G.IV bombers of *Kagohl* 3 on 13 June 1917. This shocked the world and instantly made "Gotha" a household name. The Gotha G.IV is by far the most famous and significant Gotha type.

The Gothaer Waggonfabrik AG was founded in July 1898 to manufacture railroad rolling stock and related transportation equipment. Gotha entered the aircraft business at the end of 1912 by building *Taube* monoplanes for the nearby Herzog Carl Eduard von Sachsen-Coburg flying school (later combined with *FEA* 3). Gotha also intended to compete for the German National *Flugspende* and other financial aviation prizes. The *Fliegertruppe* appreciated the sturdy structure of the Gotha *Taube* monoplanes, designed by *Diplom-Ingenieur* Karl Rösner, *Diplom-Ingenieur* Karl Grulich, Böhnisch, and Bartle, and this resulted in substantial production orders placed before the war. Gotha *Tauben* were active on various war fronts until replaced by biplanes in mid-1915.

Before the war Gotha also financed the Hanseatische Flugzeugwerke Karl Caspar and built *Taube* trainers used by the flying school. This business relationship was terminated in June 1914.

The single-engine Gotha biplanes designed by *Ingenieur* Hans Burkhard, Rösner, Schmieder, and Zimmerman were less successful. Only a few saw combat; most were flown as trainers at Herzog Carl Eduard flying school.

Gotha's significant aviation breakthrough occurred in April 1915 when it received a contract to produce the unusual Friedel-Ursinus battleplane under the designation Gotha G.I. After gaining some experience with twin-engine aircraft, Burkhard designed the Gotha bombers which, due to their range and payload, were to gain notoriety as the aircraft that bombed London. These bombers were tail heavy, making them marginally stable, demanding to fly, and difficult to land, resulting in numerous operational accidents and necessitating a variety of modifications that generally reduced performance. During the war 522 Gotha bombers were built by Gotha, LVG, and SSW. In August 1918 *Idflieg* awarded Gotha a contract for production of 50 Friedrichshafen G.IVa(Go) bombers under license, clearly indicating the most famous Gotha design was no longer competitive.

The last Gotha bomber designs were lightened twin-engine aircraft (designated GL-types) designed for high speed and high altitude. The GL.VII, GL.VIII, and GL.IX designed by Rösner and Schleiffer were promising in early flight tests but were disappointing when tested to operational requirements. Development was frustrated by lack of suitable high-altitude engines and propellers. Flight tests with supercharged engines and controllable-pitch propellers were underway when the war ended. Development of the last design, the GL.X designed by Burkhard, was terminated by the Armistice.

When Graf Zeppelin decided to manufacture giant bombers (*Riesenflugzeug*) to supplement dirigibles, the new subsidiary rented a large hangar at the Gotha factory to build the first giant bomber. Known under the cover name Versuchsbau Gotha-Ost (Experimental Works Gotha-East), VGO borrowed skilled workers and wood-working machinery from Gotha in addition to renting the hangar. The first three aircraft were designated VGO; the remaining aircraft were known as Staaken Giants.

Surprisingly, despite being far from the ocean, Gotha decided to enter the seaplane business and achieved a fair degree of success. The first Gotha seaplane design was a pusher floatplane designed by Bruno Büchner for the *Bodensee Wettbewerb* in June 1913. In late 1913 Rösner, assisted by *Ingenieur* A. Klabue, designed the conventional Gotha WD1 seaplane. This team continued to design and build seaplanes through a series of two-and four-engine floatplanes. Production for the German Navy was modest; Gotha's 177 seaplanes delivered during the war, including 32 for the Turkish Navy, rank a distant third in German seaplane deliveries behind Friedrichshafen (854) and Hansa-Brandenburg (671).

Gotha's most significant seaplane contribution was the twin-engine torpedo bombers that were superior to those by Friedrichshafen and Hansa-Brandenburg. However, extensive trials and combat sorties demonstrated that flying at very low level while arming, pre-heating, aiming, and launching a torpedo was impractical unless conditions were optimum. Furthermore, because the torpedo runs had to be made at constant heading and altitude at low speed and very low altitude, torpedo bombers were highly vulnerable to shipboard anti-aircraft gunners. As a result torpedo attacks were stopped and the twin-engine Gotha torpedo bombers were modified into long-range reconnaissance aircraft by replacing the torpedo with extra fuel tanks. The intention was to monitor British fleet movements in the North Sea, but few, if any, operational long-range sorties were performed. When inspected by the Inter-Allied Naval Commission in December 1918, most of the twin-engine Gotha floatplanes were found stored at the Hage aircraft depot.

From 1913 through December 1919 Gotha delivered 648 aircraft comprised of 112 *Taube* monoplanes, 39 single-engine biplanes, 177 seaplanes, and 325 twin-engine bombers. At the end of 1917 Gotha employed 595 workers dedicated to manufacturing aircraft. According to the Inter-Allied Aeronautical Control Commission, the total rose to 1,200 workers in 1918. On 20 January 1919 the German government ordered Gotha to stop all military aircraft activity. Gotha terminated all aviation work and did not resume building aircraft until establishment of the Third Reich.

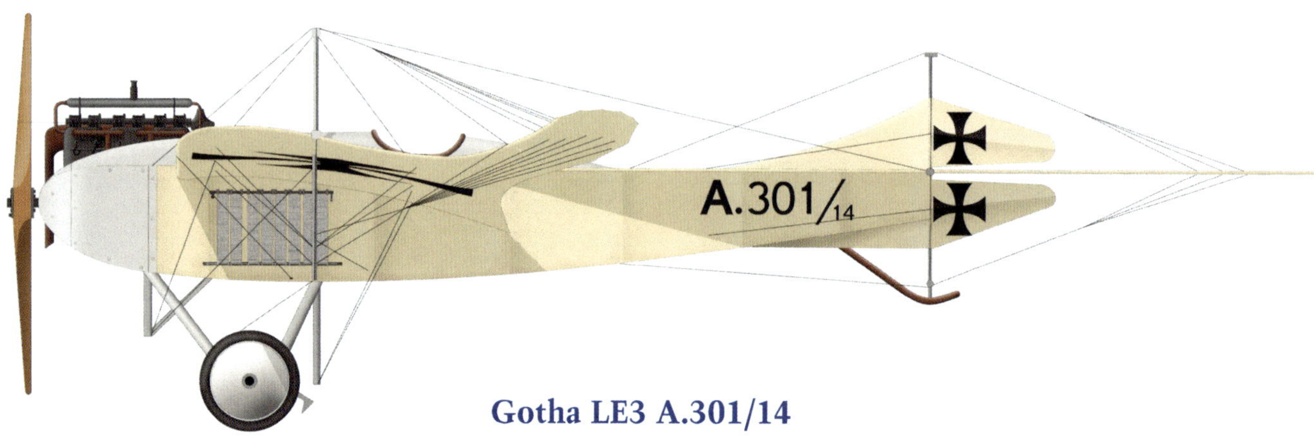

Gotha LE3 A.301/14

Gotha LE Types

Before WWI the *Taube* (Dove in German) configuration was very popular in Germany and Austria for its inherent stability, considered quite important at the time due to the limited flying experience of most airmen. Many German and Austrian manufacturers built aircraft to the *Taube* configuration and this is how Gotha got its start in aviation. The Gotha LE (*Land Eindecker* = land monoplane) series were all *Taube* designs.

Gotha's *Tauben* were rugged, well-built airplanes that saw reliable service on all war fronts. A total of 90 Gotha *Tauben* were purchased by the *Fliegertruppe* in 1913 and 1914, more than any other manufacturer except Jeannin.

Above: The Gotha LE3 Taube was the LE-type produced in greatest numbers. A robust, well-built machine, it was well suited for operational use and saw service on all fronts.

Right: The Gotha logo on a Gotha-built propeller included a Gotha *Taube* in its design.

Gotha LE1

Ordered on 4 February, 1913, eight Gotha LE1 *Taube* two-seat trainers were delivered to the Gotha flying school and two uncovered airframes were sent to Hamburg to serve as trainers at the Centrale für Aviatik, an organization in which Gotha had a financial interest. In 1914 Centrale für Aviatik was renamed Hansa Flugzeugwerke Hamburg Karl Caspar.

Designed by Gurlich and Böhnisch, the first Gotha LE1, powered by a 75 hp Mercedes engine flew its maiden flight on 22 April 1913. Two LE1 monoplanes participated in the *Prinz Heinrich Flug* of 1913; the one flown by *Lt*. Joly, with *Oblt*. Felmy as passenger, completed all the competition stages. A few LE1 trainers served into the early war months.

Gotha LE1 Specifications

Engine:	75 hp Mercedes	
Wing:	Span	14.40 m
	Area	35.2 m^2
General:	Length	7.80 m
	Height	3.20 m
	Empty Weight	600 kg
	Loaded Weight	972 kg
Maximum Speed		90 km/h
Climb:	800m	12 min

Below & Bottom: The Gotha LE1 was the first Gotha aircraft; it first flew on 22 April 1913. It was a typical *Taube* design for the period. Popular in Germany, the *Taube* configuration was basically obsolete in 1912, before the LE1 was designed. Ten LE1 aircraft were built.

Gotha LE2

Designed by Böhnisch and Bartel, the next Gotha design was the LE2, another *Taube* clearly derived from the LE1. The LE2 had a more powerful, 100 hp Mercedes D.I engine and the landing gear was redesigned to better support the wing bracing cables. The LE2 was slightly faster than the LE1 but its rate of climb was lower, likely due to a combination of greater weight and reduced wing area.

In addition to the 100 hp Mercedes D.I, LE2s were fitted with a variety of Argus, Benz, and Rapp 100 hp engines, and a few were fitted with 80 hp Oberursel U.0 engines to provide pilots with experience flying aircraft with rotary engines.

Gotha's reputation for good workmanship and solid construction led the *Fliegertruppe* to order 35 Gotha A (LE2) *Tauben*. These were used both for combat missions over the front and training.

Gotha LE2 Specifications

Engine:	100 hp Mercedes D.I	
Wing:	Span	14.40 m
	Area	28 m^2
General:	Length	10.22 m
	Height	3.20 m
	Empty Weight	690 kg
	Loaded Weight	1,053 kg
Maximum Speed		102 km/h
Climb:	800m	20 min

Above: Gotha LE2 after a landing accident.

Below: Gotha LE2 with crew Schlegel and *Lt.* Spang.

Above: Compared to the LE1, the Gotha LE2 had a redesigned landing gear and a more powerful engine. On 4 November 1914, *Lt*. Caspar and *Oblt*. Roos, flying a Gotha LE2, made history when they crossed the English Channel and dropped two bombs on Dover, returning safely after a flight of five and a half hours.

Above & Below: Front and rear quarter views of a Gotha LE2. A limitation of all *Tauben* was their sluggish wing- and tail-warping controls and automatic stability, which made them inadequate for effectively training pilots to fly comparatively maneuverable biplanes fitted with separate ailerons and elevators. Moreover, their complex rigging required painstaking adjustment and created excessive drag, limiting speed and climb. For these reasons *Tauben* were quickly superseded by biplanes with conventional controls after the war began.

Gotha LE3

The LE1 and LE2 had fuselages of triangular cross section which tended to twist in flight, so Grulich and Bartl designed the LE3 with a much stiffer fuselage of rectangular cross section. The undercarriage was again redesigned and simplified, and a separate trestle supported the lift wires in order to separate them from the undercarriage.

On 10 January 1914 work began on the LE3. In June the *Fliegertruppe* ordered 16 LE3s with the first delivery on 31 August. Upon mobilization, eight LE3s built for the Gotha flying school were commandeered by the *Fliegertruppe* for combat. In September a further 20 were ordered, followed by 10 more in October, making a total of 54 accepted by the *Fliegertruppe*. The LE3 was a robust aircraft that saw combat on all fronts. The last LE3 was delivered to FEA 3 on 7 July 1915. LE3 A.90.14 was fitted with a 160 hp Benz engine.

Gotha LE3 Specifications

Engine:	100 hp Mercedes D.I	
Wing:	Span	14.50 m
	Area	33.5 m^2
General:	Length	10.00 m
	Height	3.15 m
	Empty Weight	690 kg
	Loaded Weight	1,062 kg
Maximum Speed		96 km/h
Climb:	800m	12 min
Range:		385 km

Below: The LE3 featured a rectangular fuselage for greater stiffness to prevent twisting in flight. The iron crosses indicate combat service in the war.

Bottom: The LE3 also introduced a simplified undercarriage to separate the wing lift wires from the undercarriage structure.

Above : Gotha LE3 serial A.301/14. Many of the unarmed monoplanes used by the *Fliegertruppe*, nearly all two-seaters, were *Tauben*. The military serial and insignia confirm this LE3 is in operational service.

Below: Gotha LE3 A.300/14 parked in a hangar. The rudder of an Albatros B-type obscures the tail of the LE3. In all, Gotha built 111 *Taube* monoplanes; 89 of those were delivered to the *Fliegertruppe*.

Gotha LE4

Designed by Karl Rösner, work on the Gotha LE4 began on 31 January 1914. Powered by a 100 hp Mercedes D.I engine, the LE4 was Gotha's last *Taube* design. The nose radiator and tail surface with conventional, hinged elevators made the LE4 much more modern in appearance.

Initially the LE4 had the lift wires attached directly to the undercarriage struts, a configuration disliked by the *Fliegertruppe*. The LE4 was damaged on 15 May 1914 while participating in the *Prinz Heinrich Flug*. During repair the LE4 was modified to have the front and rear lift wires attached to separate pylons in front and behind the undercarriage. The repaired LE4 was entered in several more flying meetings before the war. After being rebuilt, the LE4 was assigned to the Herzog Carl Eduard Fliegerschule in Gotha. Only one LE4 was built.

Gotha LE4 Specifications

Engine:	100 hp Mercedes D.I	
Wing:	Span	14.00 m
	Area	28 m^2
General:	Length	8.50 m
	Height	2.80 m
	Loaded Weight	980 kg
Maximum Speed		120 km/h
Range:		600 km

Above: The Gotha LE4 after modification and assignment to the Herzog Carl Eduard Fliegerschule in Gotha.

Below: Rear view of the Gotha LE4; lack of national insignia indicates this is the original configuration before repair.

Above: The original configuration of the Gotha LE4. Gotha test pilot Oswald Kahnt, commemorated on this card, was killed when the later LD6 crashed on its maiden flight.

Below: The Gotha LE4 in flight.

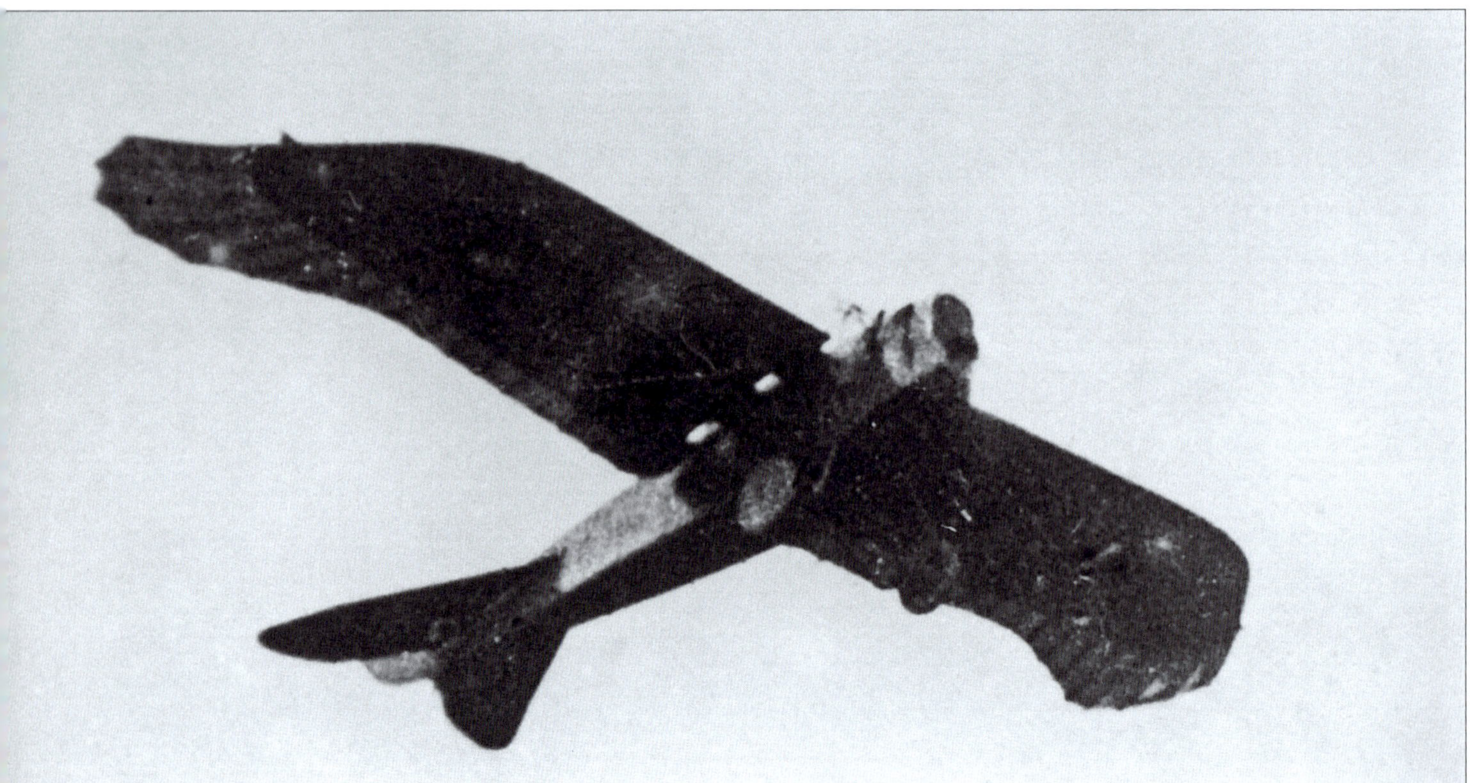

Gotha LD Types

Following the Gotha *Taube* designs, Gotha built a series of undistinguished single-engine land biplanes. Two types were license-built Caudrons whose production was apparently limited to single machines. Of the remaining five designs original to Gotha, only two were given official military designations and built in small quantities.

Gotha LD1

Designed by Rösner, the Gotha LD1 biplane (LD = *Land Doppeldecker*, or land biplane), powered by a 100 hp Mercedes engine, was ordered on 24 February 1914. It was a conventional, three-bay design with side-mounted radiators. It may have been intended as a competition machine but appears to have remained at Gotha for evaluation and training. During the course of flight trials, the LD1 received a new rudder and other minor modifications.

On 31 August 1914 the LD1 was commandeered by officers of *Feld-Flieger-Abteilung* 4 and subsequently purchased by the *Fliegertruppe*, which assigned it the designation B.458/14. Only one LD1 was built.

Gotha LD1 Specifications

Engine:	100 hp Mercedes D.I	
Wing:	Span Upper	12.55 m
	Span Lower	12.25 m
	Area	36 m^2
General:	Length	8.00 m
	Height	3.25 m
	Empty Weight	525 kg
	Loaded Weight	935 kg
Maximum Speed:		120 km/h
Climb:	800m	12 min
	2000m	30 min
Range:		450 km

Above: The Gotha LD1 was the start of Gotha's series of single-engine landplane biplanes. Powered by a 100 hp Mercedes D.I engine, it had a reasonable performance for its time but only one was built. Used at Gotha for training, it was commandeered by officers of *Feld-Flieger-Abteilung* 4 on 31 August 1914 and subsequently purchased by the *Fliegertruppe*, which assigned it the designation B.458/14. Only one LD1 was built.

Gotha LD2 (Gotha B.II)

Designed by Rösner, the Gotha LD2 was powered by a 100 hp Oberursel rotary engine. Intended to achieve maximum speed, possibly to compete for the speed prize of 300,000 Marks posted by the *National Flugspende*, work on the LD2 prototype began on 31 March 1914.

On 24 August 1914, *Idflieg* ordered nine Gotha LD2 biplanes, numbered B.459–467/14. These were later assigned the designation Gotha B.II. The production prototype had a revised cowling for better engine cooling, and the cowling was again revised for the main production run, which also featured cut-down fuselage decking. Some time during the design's evolution the fixed vertical fin was eliminated, the revised aircraft having a button rudder. Starting 12 April 1915 all nine aircraft were delivered to *FEA* 3.

Gotha LD2 Specifications

Engine:	100 hp Oberursel U.I	
Wing:	Span Upper	14.50 m
	Span Lower	13.50 m
	Area	46 m²
General:	Length	8.28 m
	Height	3.45 m
	Empty Weight	590 kg
	Loaded Weight	982 kg
Maximum Speed:		115 km/h
Climb:	1000m	12 min
	2000m	32 min
Range:		520 km

Below & Bottom: Two views of the prototype Gotha LD2 showing its streamlined engine cowling and spinner and very low profile fin and rudder.

Above: The production prototype of the Gotha LD2 was far less appealing than the first prototype. The cowling was modified for better cooling and the fin and rudder were now enlarged for better stability and control.

Below: Another view of the production prototype Gotha LD2 shows the unlovely cowling as redesigned for better cooling. National insignia were prominently painted on the wings.

Gotha LD3

Supervised by *Ingenieur* H. Schmieder, the Gotha LD3 was powered by a 50 hp Gnome rotary engine. The LD3 was a license-built Caudron. Work started on the LD3 on 7 October 1913; upon completion the aircraft was assigned to the Herzog Carl Eduard Fliegerschule. Only one LD3 was built.

Gotha LD3 Specifications

Engine:	50 hp Gnôme	
Wing:	Span Upper	13.40 m
	Span Lower	10.05 m
	Area	26.4 m^2
General:	Length	6.40 m
	Height	2.5 m
	Empty Weight	340 kg
	Loaded Weight	630 kg
Maximum Speed:		110 km/h

Below: The Gotha LD3 was a license-built Caudron; only one was built.

Left: The Gotha LD3 after an engine failure.
Above: Close-up of the engine after failure in the LD3.

Gotha LD4

The Gotha LD4, powered by a 100 hp Gnome rotary engine, was another license-built Caudron but with side-by-side seating. It was ordered on 26 February 1914. Only one LD4 was built.

Below & Bottom: The Gotha LD4, a license-built Caudron, had a spinner and streamlined nose similar to that of the prototype LD2.

Gotha LD4 Specifications		
Engine:	100 hp Gnôme	
Wing:	Span Upper	13.40 m
	Span Lower	10.05 m
	Area	27 m^2
General:	Length	6.40 m
	Height	2.5 m
	Empty Weight	420 kg
	Loaded Weight	710 kg

Gotha LD5

On 17 October 1914 Gotha began work on the LD5, which was designed by *Ingenieur* Hans Burkhard, a Swiss citizen who worked for Gotha throughout the war. Prior to joining Gotha in 1914, Burkhard had designed aircraft for Rumpler (1911), Bristol (1912), and Halberstadt (1913).

Powered by a 100 hp Oberursel rotary engine, the Gotha LD5 was a "cavalry biplane" intended as a light spotting aircraft to accompany an army on the move. The LD5, which first flew in December 1914, was a small, two-bay biplane with no fixed tail surfaces. Never assigned an official designation, the sole Gotha LD5 later served as a trainer at *FEA* 3.

Gotha LD5 Specifications	
Engine:	100 hp Oberursel U.I

Above & Below: The Gotha LD5, powered by a 100 hp Oberursel U.I, was designed as a 'cavalry biplane', a light observation plane designed to accompany an army on the move. The sole LD5 built served as a trainer at *FEA* 3.

Gotha LD6

Construction of the Gotha LD6 began on 17 December 1914. Powered by a 150 hp Benz Bz.III engine, the LD6 was a larger, more powerful aircraft than its predecessors. Unfortunately, Gotha test pilot Oswald Kahnt was killed when the LD6 crashed on its maiden flight on 30 January 1915.

Undeterred by this setback, *Idflieg* ordered the modified Gotha LD6a with larger tail surfaces and other improvements in February 1915. Failing to meet the acceptance specifications, the LD6a was assigned to the Herzog Carl Eduard Fliegerschule, where it received school number 70.

Gotha LD6 Specifications	
Engine:	150 hp Benz Bz.III

Above & Below: The ill-fated original LD6 with small fin and rudder. It crashed on its maiden flight, killing test pilot Oswald Kahnt.

Above: Another view of the original LD6 with small fin and rudder that crashed on its first flight.

Below: Side view of the revised LD6a with larger fin and rudder. It failed to meet the *Idflieg* acceptance requirements and was assigned to the Herzog Carl Eduard Fliegerschule, where it received school number 70.

Gotha LD7 (Gotha B.I)

Designed by *Ingenieur* Hans Burkhard, Gotha began work on the LD7 on 10 December 1914. Powered by a 100 hp Mercedes, the LD7 was intended as a trainer for the Herzog Carl Eduard Fliegerschule. In May 1915 *Idflieg* ordered 18 LD7 trainers under the designation Gotha B.I. These were numbered B.960–974/14 and B.884–886/15. Early delivery was a priority, but modifications for observation missions and a switch from the 100 hp to the 120 hp Mercedes substantially delayed production, and the last B.I was not accepted until 11 November 1915.

Five machines, B.961–965/14, were dispatched to the German Air Service in Turkey via Chernowitz in August–September 1915. The remaining aircraft were used as trainers at *FEA* 3 in Gotha.

Gotha LD7 / B.I Specifications

Engine:	120 hp Mercedes D.II	
Wing:	Span Upper	12.40 m
	Span Lower	12.07 m
	Area	39.5 m^2
General:	Length	8.40 m
	Height	3.00 m
	Empty Weight	725 kg
	Loaded Weight	1125 kg
Maximum Speed:		125 km/h
Climb:	800m	8.5 min
	2700m	45 min
Range:		530 km

Above: This Gotha B.I is in service at the Gotha Fliegerschule, as indicated by the circular insignia and the lettering below it. The phone number is given for the convenience of anyone finding one of these aircraft force-landed away from the aerodrome. The Gotha B.I in this photograph had a different design exhaust manifold than the one at the bottom of the next page.

Above: This Gotha B.I and aviator appear to be the same as in the previous photograph.

Below: The Gotha B.I was a conventional, two-bay biplane powered by a 120 hp Mercedes D.II engine.

Above: Gotha B.I with rhino-horn exhaust that discharged the hot, toxic gases over the wing for benefit of the crew.
Below: This side view of the Gotha B.I emphasizes the enlarged fin and rudder compared to earlier Gotha types.

Above: Gotha B.I B.961/14 is one of five sent to Turkey for use by the German Air Service. Despite being attached to the German Air Service, it carries Turkish markings.

Below: This Gotha B.I is likely in training service at *FEA* 3 at Gotha. Of the 18 aircraft built, five went to the German Air Service in Turkey and the other 13 were assigned to *FEA* 3.

Gotha WD Types

Gotha achieved a modest degree of success in the seaplane business. The first Gotha seaplane design was a pusher floatplane designed by Bruno Büchner for the *Bodensee Wettbewerb* in June 1913. In late 1913 Rösner, assisted by *Ingenieur* A. Klabue, designed the conventional Gotha WD1 seaplane. WD likely stood for *Wasser Doppeldecker*, or water biplane. This team continued to design and build seaplanes through a series of two-and four-engine floatplanes. Production for the German Navy was modest; Gotha's 177 seaplanes delivered during the war, including 32 for the Turkish Navy, rank a distant third in German seaplane deliveries.

Gotha's most significant seaplane contribution was its twin-engine torpedo bombers that were superior to those by Friedrichshafen and Hansa-Brandenburg. Unfortunately, most of these designs were tail-heavy, a problem they apparently inherited from contemporary Gotha land-plane bomber designs as first implemented on the Gotha G.II production aircraft. Because the reason for making the Gotha G.II and subsequent designs tail heavy was to enable them to more easily dig in their tailskid for braking during landings, carrying this over to seaplanes, which had no problem stopping after landing, makes no sense and was a fundamental design flaw. The larger seaplanes were also under-powered, likely as a result of Germany's desperate resource shortages due to the Allied distant blockade.

Interestingly, despite the German Navy standardizing seaplane camouflage in April 1917, Gotha floatplanes apparently did not fully conform to the standard. Most available photos of the WD11, which usually do not show the upper surfaces clearly, appear to show a light, plain finish instead of three-color camouflage fabric on upper surfaces. Production WD14s and later seaplanes wear dark camouflage fabric overall instead of on just the upper surfaces.

Above: The Büchner *See* pusher floatplane was the first Gotha seaplane. Built in 1913, it was powered by a 100 hp Mercedes D.I. Only one was built, but it started Gotha in the seaplane business.

Gotha Büchner *See* Specifications

Engine:	100 hp Mercedes D.I	
Wing:	Span Upper	20.00 m
General:	Length	10.50 m
	Height	3.90 m
	Empty Weight	750 kg

Gotha WD1

The first Gotha WD1 was a private venture completed in February 1914 and powered by a 14-cylinder, 100 hp Gnôme rotary. The second WD1 powered by a 100 hp Mercedes D.I was built for the *Ostseeflug Warnemünde* 1914. This aircraft, competition number 20, was purchased by the Navy on 8 August 1914 and assigned Marine Number 59.

On 14 December 1914 the Navy purchased five Gotha WD1 seaplanes, Marine Numbers 285–289. Powered by a 100 hp Mercedes, these were delivered between 17 February and 24 March 1915. The WD1 suffered from poor aileron control and long take-off runs.

Despite being unarmed and low-powered, three WD1 seaplanes, Marine Numbers 286, 287, and 289, were sent to the German *Wasserfliegerabteilung* in Turkey, arriving in July 1915. These seaplanes were used for reconnaissance and light bombing of Allied forces on Imbros and the Gallipoli peninsula. These seaplanes also conducted anti-submarine patrols over the Sea of Marmara and attacked British submarines on several occasions. In October 1915 three armed Gotha WD2 seaplanes were added to the German unit. On 29 Nov. 1915 both #287 and #289 were lost in separate incidents due to mechanical failure.

Gotha WD1 Specifications

Engine:	100 hp Mercedes D.I	
Wing:	Span Upper	14.10 m
	Span Lower	13.63 m
	Area	50 m^2
General:	Length	10.32 m
	Height	4.00 m
	Empty Weight	800 kg
	Loaded Weight	1220 kg
Maximum Speed:		90 km/h
Climb:	1000m	24.5 min
Service Ceiling		2500 m
Range:		540 km

Above: The prototype WD1 was powered by a two-row, 100 hp Gnôme rotary engine. The WD1 floatplane was a conventional 3-bay biplane; a small float was attached to the rudder.

Above : The six subsequent WD1 floatplanes were powered by a 100 hp Mercedes D.I engine. This WD1 is almost certainly the second WD1 built by Gotha for the *Ostseeflug Warnemünde* 1914 competition. It has the Gotha name painted on the fuselage and retains the small float attached to the rudder. This aircraft was later purchased by the Navy.

Below: This WD1 was in naval service as indicated by the national insignia and pennants attached to the outer struts. One of the series Marine Numbers 285–289, no float was attached to the rudder. Marine #285 was at Chanak seaplane station in Turkey by December 1915, while #286 was at Kawak.

Gotha WD2

Powered by a 150 hp Rapp engine, the Gotha WD2 was built for the *Ostseeflug Warnemünde* 1914. This aircraft, competition number 19a, was confiscated by the Navy at the beginning of the war and assigned Marine Number 61. Damaged several times during flight testing, it was repaired and finally accepted on 10 December 1914.

A second WD2, also entered in the *Ostseeflug Warnemünde* 1914, was competition number 19. Powered by a 150 hp Benz Bz.III, this WD2 was accepted by the Navy on 5 August 1914 and assigned Marine Number 60.

The first WD2 production series, five floatplanes with Marine Numbers 236–240 and powered by the 160 hp Mercedes D.III, was ordered in July 1914 and delivered between December 1914 and April 1915. The second WD2 production series, again of five aircraft, was given Marine Numbers 254–258. This series was ordered in March 1915, and the last aircraft was delivered that December. On 3 July 1915, accompanied by an Albatros floatplane, WD2 #257 from Zeebrügge bombed the Landguard Point signals station near Harwich without effect.

Two 160 hp Mercedes-powered WD2 aircraft were built as Marine Numbers 424–425; these were fitted with temporary wheels and a tailskid attached to each float for overland flights to Turkey because shipments through Romania were banned. These delivery flights took place in early 1916. These WD2s were fitted with a machine gun turret above the upper wing, giving the gunner a 360° field of fire.

Six WD2s were ordered for Turkey on 6 January 1916 and delivered in May–July 1916; these were powered by the 160 hp Mercedes D.III. Because these were ordered for Turkey no Marine Numbers were assigned. Five similar floatplanes were delivered to the German Navy in February–March 1916. Because the Navy critized the WD2 floatplanes for poor seaworthiness, many WD2s were sent to the Dardanelles where the weather and sea conditions were less demanding.

The last two WD2s were ordered by Turkey on 29 April 1916 and delivered in August; no Marine Numbers were assigned. These were modified by having a conventional gun ring for the observer, split radiators mounted on the center-section struts, and a large gravity tank mounted away from the engine. Turkish sources label these machines as WD13s.

Gotha WD2 Specifications

Engine:	150 hp Benz Bz.III	
	150 hp Rapp Rp.III	
	160 hp Mercedes D.III	
Wing:	Span Upper	16.43 m
	Span Lower	15.57 m
	Area	58.5 m²
General:	Length	10.50 m
	Height	4.10 m
	Empty Weight	1050 kg
	Loaded Weight	1487 kg
Maximum Speed:		96 km/h
Climb:	1000m	12 min
	2000m	32 min
Service Ceiling		3200 m
Range:		670 km

Above & Below: WD2 Marine Number 236 is seen taxiing for take-off. The float attached to the rudder of the early WD1aircraft was no longer used on subsequent types. Its 160 hp Mercedes D.III gave it much more power than its WD1 predecessor. At least three WD2 floatplanes were sent to the German *Wasserfliegerabteilung* in Turkey in October 1915. In addition, at least ten WD2 floatplanes were delivered to Turkish forces.

Facing Page: The first WD2 floatplane was powered by a 150 hp Rapp engine and was entered in the *Ostseeflug Warnemünde* 1914 competition, competition #19a. It has the Gotha name painted on the fuselage; this aircraft was later purchased by the Navy and assigned Marine Number 61.

Above & Below: WD2 Marine Number 424 (shown here) and 425 were fitted with temporary wheels and tail skids attached to their floats to enable overland flights to Turkey because rail shipments through Romania were banned. These two aircraft were powered by the 160 hp Mercedes D.III engine and had a gun turret for the observer mounted above the upper wing, giving the gunner a 360° field of fire. WD2 #424 was at Chanak when credited with a victory in Jan. 1916 by *Flgobmt* Wilhelm Schubert (pilot)/*Flgmt* Werdier (observer), who shot down a Farman that crashed east of Tenedos. Later this aircraft was at Kawak.

Above: This WD2 floatplane on wheels appears different than Marine Number 424; it does not have the gun turret and the exhaust is different. It wears German markings.

Above: This WD2 floatplane on wheels wears Turkish markings. It lacks the gun turret applied to Marine Number 424 and many other WD2 floatplanes destined for Turkish service. The gun mount for the observer can just be seen.

Above: WD2 floatplane in Turkish markings taking off. Like many WD2 floatplanes in Turkish service, it has the gun turret.

Above: Turkish WD2 floatplane.

Above: WD2 floatplane in Turkish markings in the Gotha factory pond. It has the gun turret applied to many WD2 floatplanes destined for Turkish service.

Above: Turkish WD2 floatplane in flight. It has the gun turret applied to many WD2 floatplanes destined for Turkish service.

Above: WD2 Marine Number 236 is seen in flight.

Facing Page: Detailed view of the gun turrent installation on a WD2.

Below: WD2 in the Gotha factory pond with 28 men on it for a test load.

Gotha

Gotha WD3

Powered by a 160 hp Mercedes D.III engine, the Gotha WD3 was one of the few pusher aircraft built in Germany. This aircraft, Marine Number 259, was a three-seat reconnaissance floatplane. Ordered in August 1914, it was not delivered until 14 September 1915, so was not a priority project.

The pusher configuration had advantages; the observer(s) had a clear view and field of fire forward, free of propeller turbulence and exhaust fumes. However, the additional weight and drag of the pusher design typically reduced its performance compared to tractor designs. The one WD3 built was flown at Warnemünde as a trainer.

Gotha WD3 Specifications

Engine:	160 hp Mercedes D.III	
Wing:	Span Upper	15.65 m
	Span Lower	14.35 m
	Area	54 m^2
General:	Empty Weight	1185 kg
	Loaded Weight	1729 kg
Maximum Speed:		100 km/h
Climb:	1000m	24 min
Service Ceiling		2200 m
Range:		670 km

Above: The Gotha WD3 is seen in the Gotha factory pond before its Marine Number was applied.

Below: The sole Gotha WD3, Marine Number 259, is seen in flight. It later went to Stralsund/Wiek as a training aircraft, being flown there on 2 August 1916.

Above: The WD3 floatplane in the Gotha factory pond before the Marine Number was applied.

Below: The WD3 floatplane, Marine Number 259, taxiing on the water. There are three crew members, the pilot aft and two observers forward, an unusual arrangement; most similar aircraft had only one observer.

Gotha WD4 (UWD)

The Gotha WD4, more commonly known as the UWD (for *Ursinus Wasser Doppeldecker* = Ursinus Water Biplane), had an interesting history. It was a floatplane derivative of the Gotha G.I bomber, which was actually designed by Oskar Ursinus, a civil engineer who was founder and editor of *Flugsport* magazine and not employed by Gotha. After a prototype was built, Gotha built additional aircraft as the Gotha G.I (which see).

On 14 April 1915, only two weeks after the first order for the Gotha G.I *Kampfflugzeug* was received, the Navy ordered a floatplane version, the UWD, for combat evaluation. Powered by two 160 hp Mercedes D.III engines, the UWD was delivered to the SVK (*Seeflugzeug Versuchs Kommando* = Seaplane Testing Command) on 30 December 1915. Flight testing began on 5 January 1916 and the UWD was accepted on 6 February.

Assigned Marine Number 120, the UWD arrived at Zeebrügge on 18 February 1916. It was easy to fly and had an endurance of at least 4.5 hours, enabling it to undertake long-range scouting and bombing missions. Nose-overs were not a problem like they were with the G.I landplane version.

The UWD's configuration had advantages; the observer(s) had a clear view and field of fire forward, free of propeller turbulence and exhaust fumes, and the closely-spaced engines minimized asymmetric thrust in case of an engine failure. Starting 10 March 1916 the UWD made several bombing raids along the English coast until it was damaged on 10 July while taxiing for take-off. It was struck from inventory by the Navy on 2 October 1916.

Gotha WD4 / UWD Specifications

Engines:	2 x 160 hp Mercedes D.III	
Wing:	Span Upper	20.10 m
	Span Lower	19.00 m
	Chord (Upper & Lower)	2.20 m
	Gap	2.00 m
	Sweepback	10 degrees
	Area	82 m²
General:	Length	14.20 m
	Height	4.40 m
	Empty Weight	1940 kg
	Loaded Weight	2552 kg
Maximum Speed:		138 km/h
Climb:	1000m	8.5 min
	2000m	20.5 min
	3000m	45 min
Service Ceiling:		4000 m

Above: The Gotha UWD at the factory prior to shipment to the SVK at Warnemünde on 30 December 1915.

Above: The Gotha UWD attracted a crowd during its sea trials in February 1916. So many people rode in it that it was nicknamed the "Trojan Horse".

Below: The Gotha UWD in flight displays its distinctive lines and the clear field of fire for its gunners, especially the nose gunner. The second gunner sat in the middle with the pilot aft as a result of its 'battle-plane' heritage from the Gotha G.I.

Above: The Gotha UWD shows more of its distinctive profile in this dramatic image. Interestingly, the Iron Cross national insignia was painted on a white background on the tops and undersides of both wings and both sides of the rudders. The protrusion below the nose was a fairing for dropping bombs added after the UWD was built.

Below: The Gotha UWD resting on a beaching dolly before the bomb-dropping fairing was added.

Facing Page: The *SVK* drawing of the UWD.

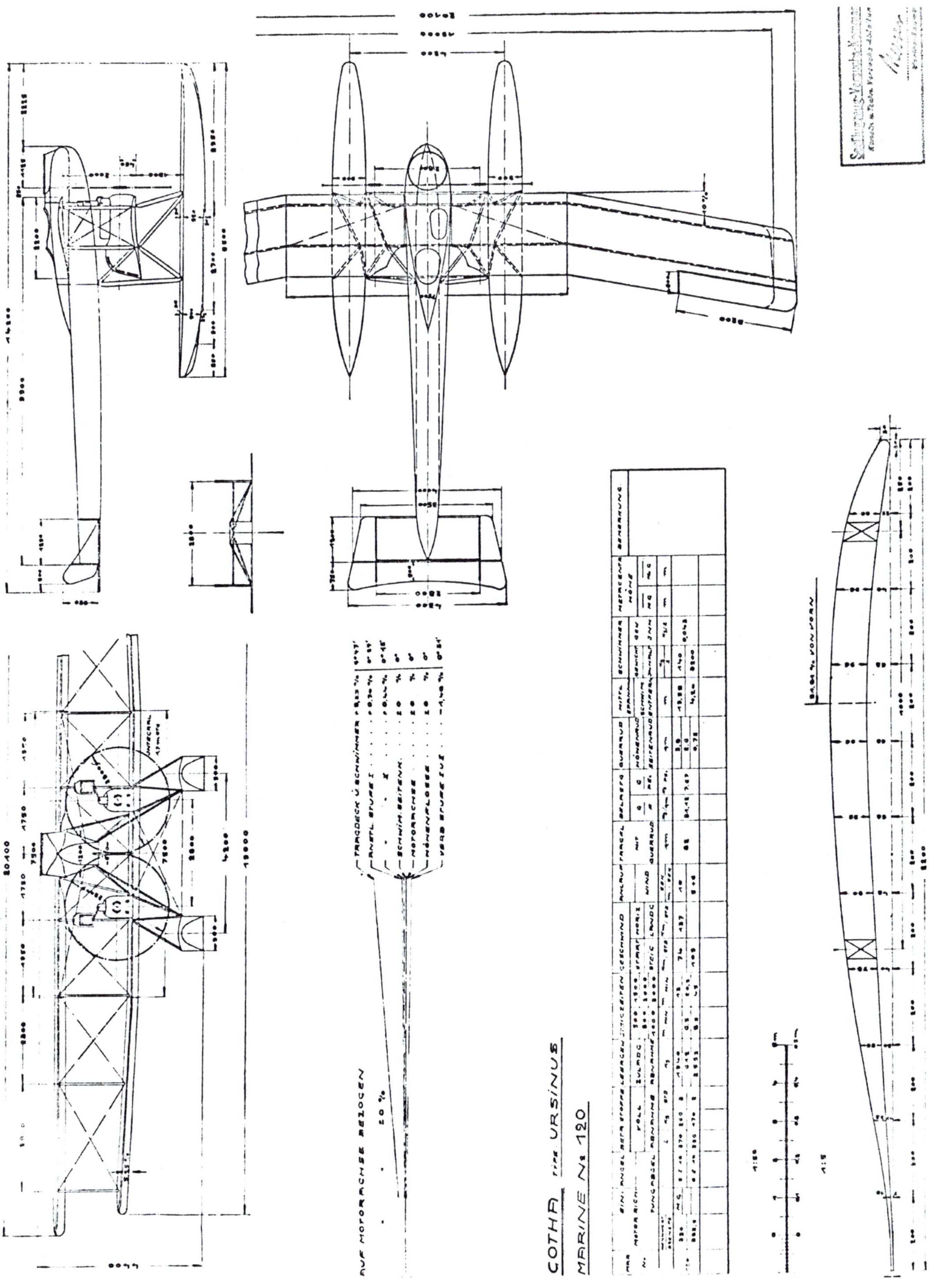
GOTHA TYPE URSINUS
MARINE No 120

Gotha WD5

The Gotha WD5, Marine Number 118, was ordered on 26 April 1915 as a high-speed seaplane for bombing surface targets. Powered by a 160 hp Mercedes D.III engine, it was the first Gotha seaplane with a two-bay wing cellule.

The WD5 was considered fast for the time but with "overly sensitive rudder" making bomb aiming difficult. Further development was not recommended, and the sole WD5 was sent to Turkey on 13 July 1916.

Gotha WD5 Specifications

Engine:	160 hp Mercedes D.III	
Wing:	Span Upper	12.00 m
	Span Lower	10.00 m
	Area	42 m^2
General:	Length	10.35 m
	Height	3.80 m
	Empty Weight	980 kg
	Loaded Weight	1465 kg
Maximum Speed:		126 km/h
Climb:	1000m	12 min
	2000m	40 min
Service Ceiling:		2800 m
Range:		440 km

Above: The Gotha WD5 in the Gotha factory pond. The compact, 2-bay design was for more speed; power was from a 160 hp Mercedes D.III engine. The insignia are on both top and bottom surfaces of the top wing.

Right: The Gotha WD5 taxiing with identification pennants attached.

Above: The Gotha WD5 in the Gotha factory pond shows its comparatively compact, 2-bay design.

Below: The Gotha WD5 on the water. Designed as a fast seaplane for bombing surface targets, the rudder was too sensitive for a steady bombing platform, and only one WD5 was built. On 13 July 1916 it was sent to Turkey.

Gotha WD6

Gotha WD6 was the Gotha company designation for 30 Hansa-Brandenburg NW(Go) seaplanes, Marine Numbers 752–782, built under license. The production order was dated 14 May 1916 and deliveries took place between 29 September 1916 and 21 February 1917. The NW was an unarmed reconnaissance floatplane that carried a radio. Some were modified to carry ten 5-kg bombs.

Gotha WD6 / Brandenburg NW(Go) Specifications

Engine:	160 hp Mercedes D.III	
Wing:	Span Upper	16.50 m
	Area	57.85 m^2
General:	Length	9.4 m
	Empty Weight	1020 kg
	Loaded Weight	1650 kg
Maximum Speed:		90 km/h
Duration:		4 hours

Above: Brandenburg NW(Go) Marine #763 getting a lift. This seaplane was destroyed at Putzig on 12 August 1918, when a student pilot was killed in a flying accident. Was this the event pictured?

Left: Brandenburg NW(Go) Marine #759 was destroyed landing at Libau on 15 June 1917.

Gotha WD7

The Gotha WD7, Marine Number 119, was ordered on 10 May 1915. Designed by Karl Rösner and A. Klaube and originally intended as a three-seat *Kampfflugzeug* to attack enemy aircraft, it was powered by two 120 hp Mercedes D.II engines. Assigned to *Flandern* I at Zeebrügge, the WD7 prototype caught fire in flight on its first combat mission on 5 April 1916 and was forced to ditch. The crew and partly-burned airframe were captured by the French.

Seven additional WD7 floatplanes were ordered on 24 February 1916. Six, Marine Numbers 670–675, were powered by two 100 hp Mercedes D.I engines and one, Marine Number 676, was powered by two 120 hp Argus As.II engines. Armament was a single Parabellum LMG 14 machine gun in the front turret. These provided training at Apenrade for twin-engine torpedo airplanes. WD7 #671 was at Warnemünde, then went to Flensburg. WD7 #674 eventually was at Norderney, where among its pilots was *Flgobmt*. Hubrich, a future ace. Two WD7 seaplanes, #675 and #676, were used in weapons trials with 37mm DWN and 20mm Becker aircraft cannons mounted in a modified nose turret.

Gotha WD7 Specifications

Engines:	2 x 120 hp Mercedes D.II (119)	
	2 x 100 hp Mercedes D.I (670–675)	
	2 x 120 hp Argus As.II (676)	
Wing:	Span Upper	16.80 m
	Span Lower	14.80 m
	Area	55.5 m^2
General:	Length	11.30 m
	Height	3.90 m
	Empty Weight	1275 kg
	Loaded Weight	1785 kg
Maximum Speed:		128 km/h
Climb:	1000m	9.5 min
	2000m	40 min
Service Ceiling:		3500 m
Range:		475 km

Note: Performance specs for #119

Above: The prototype Gotha WD7, Marine #119, in the Gotha factory pond. The WD7 had distinctive tail surfaces with three fins and two rudders with aerodynamic balances at the bottom. The WD7 was designed as a 3-seat *Kampfflugzeug* meant to destroy enemy aircraft, but by the time the production batch was available that concept was out-moded.

Above: The Gotha WD7 prototype afloat.

Below: On its first combat mission on 5 April 1916, WD7 Marine Number 119 caught fire in flight and was forced to make an emergency landing. Here it is photographed from a French vessel that captured the crew and partly-burnt airframe.

Right: Gotha WD7 #119 after capture. The WD7 was indeed "brought down in flames off the coast of Dunkirk" as written on the photograph but that was due to mechanical failure, not Allied action. WD7 #119 was the subject of a French report in *L'Aerophile* from where we learn it was doped a very light blue.

Above: Gotha WD7 Marine Number 671 afloat in the Gotha factory pond.

Below: Side view of Gotha WD7 Marine Number 674 on a dolley. Power for these aircraft was two 100 hp Mercedes D.Is.

Gotha WD8

The Gotha WD8, Marine Number 476, was ordered on 22 July 1915. Designed by Karl Rösner and A. Klaube and powered by a single 240 hp Maybach Mb.IVa, it was ordered as a "comparison aircraft" for the twin-engined WD7 which had the same horsepower. The WD8 used the same wing cellule, tail assembly, and floats as the WD7, and was armed with both a fixed, forward-firing machine gun and a flexible rear machine gun for the observer.

Delivered on 9 February 1916, the WD8 was assigned to *Flandern* I at Zeebrügge, where it arrived on 23 April 1916. On operations the WD8 was reported as "totally unsuitable" owing to sluggish maneuverability, insufficent fuel capacity, and a poor field of fire due to the twin rudders. On July 28 the commander of the *Seeflugstation Zeebrügge* requested that the WD8 be withdrawn from front-line service. The WD8 was shipped to Turkey by rail on 21 September 1916, and apparently arived on 31 October. Only one WD8 was built.

Gotha WD8 Specifications

Engine:	240 hp Maybach Mb.IVa	
Wing:	Span Upper	16.00 m
	Span Lower	14.80 m
	Area	55.5 m^2
General:	Length	11.20 m
	Height	4.10 m
	Empty Weight	1254 kg
	Loaded Weight	1778 kg
Maximum Speed:		138 km/h
Climb:	1000m	6.5 min
	2500m	25 min
Service Ceiling:		4500 m
Range:		480 km

Above: Gotha WD8 Marine Number 476 afloat in the Gotha factory pond. The WD8 used the same wing cellule, tail surfaces, and floats as the WD7 and was built as a comparison aircraft to the WD7 because both aircraft had the same power. Confusingly, the wrong Marine Number was painted on the floatplane when these photos were taken, a mistake later corrected.

Above: Side view of Gotha WD8 Marine Number 476 afloat in the Gotha factory pond. The wrong Marine Number was painted on the WD8 when these photos were taken. With the same total engine power as the WD7 had from two engines, the WD8 was 10 kmh faster due to lower drag.

Above: Front view of Gotha WD8 Marine Number 476 in the Gotha factory pond. The sole WD8 was assigned to Zeebrügge, where it was judged "totally unsuitable" for operations due to sluggish maneuverability, a poor field of fire for the rear gunner due to the large, complex tail surfaces, and insufficient fuel capacity. After a few months of operations the commander of *Seeflugstation Zeebrügge* requested the WD8 be removed from front-line service.

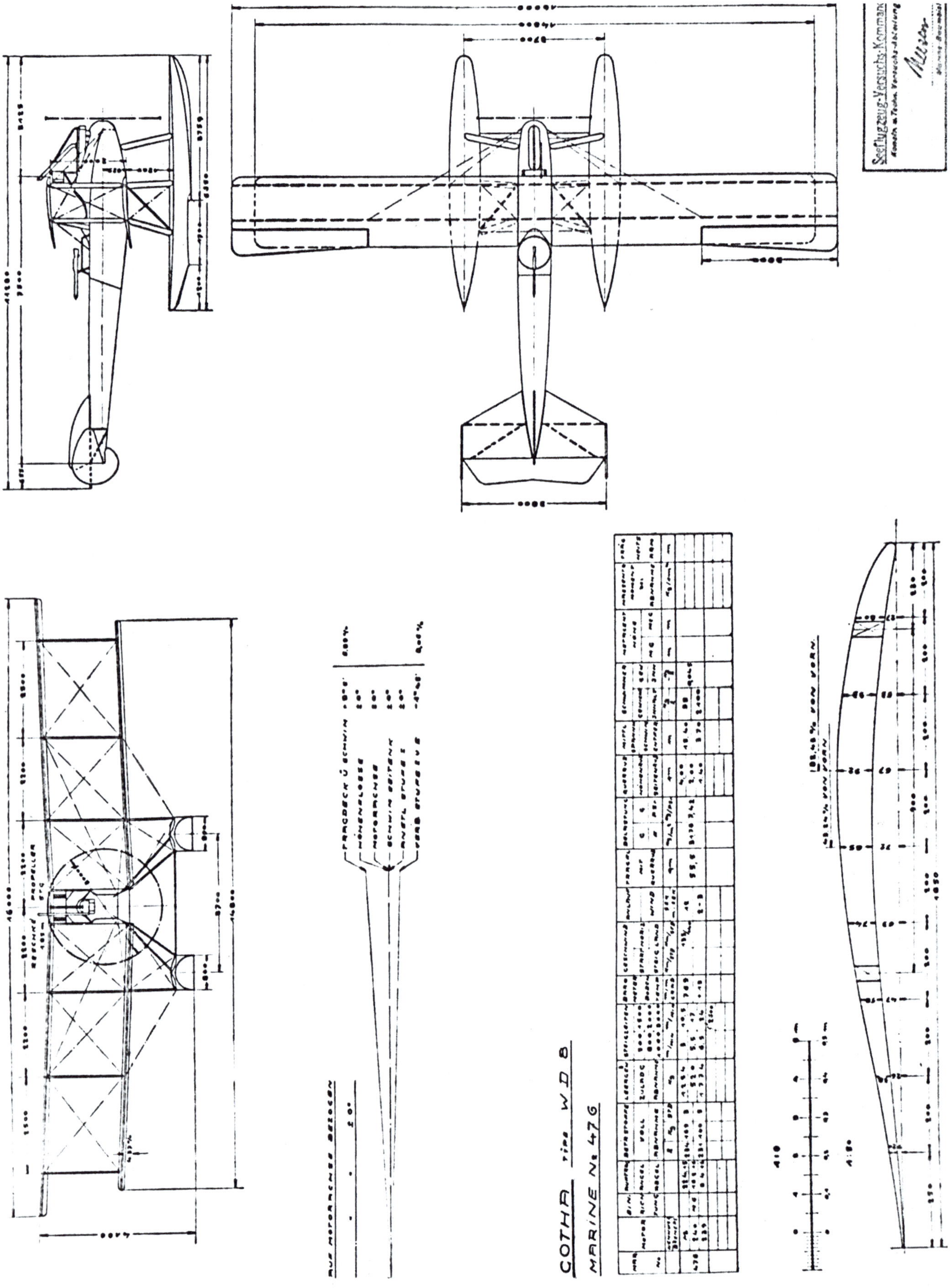

Gotha WD9

The Gotha WD9, Marine Number 572, was ordered on 26 October 1915. Designed by Karl Rösner and A. Klaube and powered by a 160 hp Mercedes D.III, it was derived from the WD5. The WD9 had a flexible gun for the observer; to increase the field of fire, the inner wing bay bracing was eliminated to enable the gunner to fire forward between the wings.

Delivered on 19 April 1916, the WD9 was assigned to *Flandern* I at Zeebrügge, where it arrived 10 May 1916. It remained until June 1916. It was ready to be sent to Turkey on 27 Sept 1916. It eventually was at Chanak, where it was flown by several aviators, including *Vflgmr*. Schubert.

Gotha WD9 Specifications

Engine:	160 hp Mercedes D.III	
Wing:	Span Upper	12.50 m
	Span Lower	11.30 m
	Area	42.5 m^2
General:	Length	9.80 m
	Height	3.80 m
	Empty Weight	967 kg
	Loaded Weight	1472 kg
Maximum Speed:		132 km/h
Climb:	1000m	12.5 min
	2000m	36 min
Service Ceiling:		3000 m

Above: Gotha WD9 Marine Number 572 on beaching dollies. The WD9 was derived from the earlier WD5 and used the same 160 hp Mercedes D.III engine, but unlike the WD5 it was armed with a flexible gun for the observer. The WD9 had an enlarged fin and rudder to improve directional stability, and aerodynamic refinements made it slightly faster.

Zeebrügge, Zeebrugge, or Seebrügge?

The German spelling was *Seebrügge*, while the Flemish spelling was *Zeebrugge,* with no umlaut. The front covers of all the WWI German files and documents on that station bear the former name, and the Germans consistently spelled it that way through most of the war. Only toward the end of the war did it sometimes appear as "*Zeebrügge*" in the German documents, apparently as an occasional misspelling combining the Flemish beginning with the German ending. *Zeebrügge* connotes WWI to us.

Above: Gotha WD9 Marine Number 572 on beaching dollies. The inner wing bay wire bracing of the WD9 was eliminated to enable the gunner to safetly fire forward between the wings. Only one WD9 was built. Initially assigned to *Flandern* I at Zeebrügge, it was later sent to Turkey.

Below: Gotha WD9 Marine Number 572 on beaching dollies. The WD9 had an aerodynamic balance on the bottom of the rudder, a feature that by now was becoming a hallmark of Gotha floatplanes.

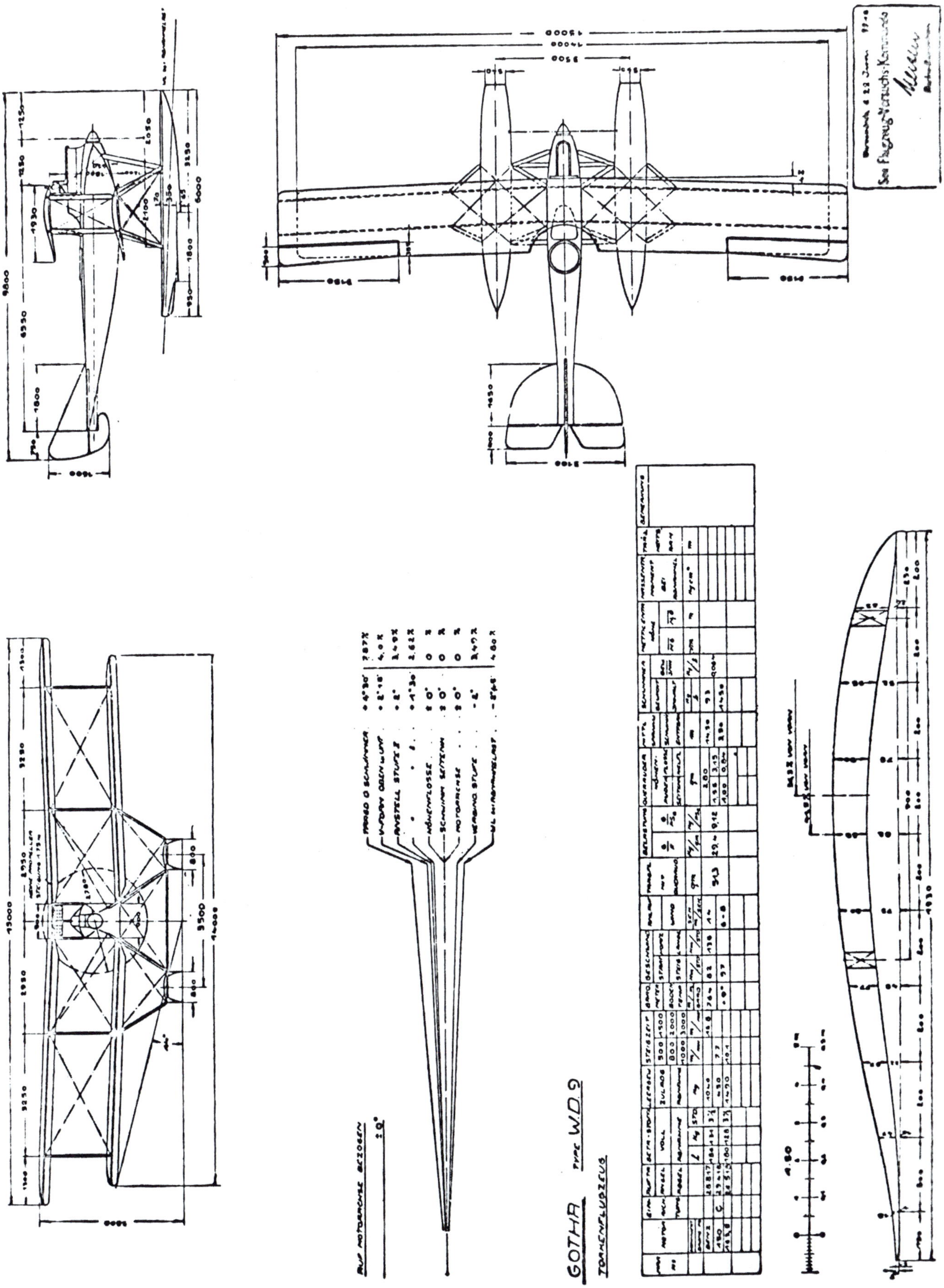
GOTHA type W.D.9
TORPEDOFLUGZEUG

Gotha WD10 (U-1)

Like the Gotha WD4/UWD, the Gotha WD10, Marine Number 782, was another design by Oskar Ursinus, a civil engineer who was founder and editor of *Flugsport* magazine.

The WD10, more commonly known as the U-1 or Ursinus Seaplane after its designer, was the most innovative single-seat floatplane fighter prototype built in Germany and a remarkable design for 1916. It featured retractable floats for higher speed through reduced drag. To improve maneuverability the 150 hp Benz Bz.III engine was located on the center of gravity and drove the propeller via an extension shaft. The floats were retracted by a manual crank.

Gotha WD10 (U-1) Specifications

Engine:	150 hp Benz Bz.III	
Wing:	Span Upper	9.00 m
	Area	42.5 m^2
General:	Length	7.77 m
	Height	2.90 m
	Empty Weight	750 kg
	Loaded Weight	1000 kg
Maximum Speed:		200 km/h

During trials there were problems with the propeller extension shaft and float retraction mechanism. Unfortunately, the prototype was destroyed before it could achieve its estimated top speed of 200 km/h.

Above & Below: The Ursinus floatplane fighter was unique in having retractable floats, a significant innovation for its time. The inside upper edges of the floats were beveled to fit closely against the fuselage, which was shaped to accommodate them when retracted.

Above: The Ursinus floatplane fighter, Marine Number 782, seen from the side with floats extended. The cockpit was well aft because the engine was set back to the center of gravity for improved maneuverability (due to reduced moment of inertia) and drove the propeller via an extension shaft.

Above: The Ursinus was an innovative design attempt to reduce the drag penalty of floats combined with an attempt to improve maneuverability through reducing the moment of inertia by locating the engine at the center of gravity.

Above: The Ursinus floatplane fighter, Marine Number 782, seen from the side with floats extended. The mechanical retraction mechanism was operated by the pilot. Unsurprisingly, the two significant mechanical innovations, the propeller extension shaft and float retraction mechanism, gave problems during testing. Only one aircraft was built.

Gotha WD11

Ordered on 9 March 1916 and designed by Rösner as a torpedo bomber, the WD11 resembled the earlier WD7 but was a much larger aircraft. The WD11 was powered by two 160 hp Mercedes D.III engines and was equipped to carry one Whitehead G/125 torpedo under the fuselage. All WD11 torpedo bombers were flown by a two-man crew, and the observer in the front cockpit had a flexible Parabellum LMG 14 machine gun.

After flight tests at the *SVK*, the prototype WD11, Marine Number 679, was used to train aircrews at the *Sonderkommando Flensburg*. Three more production batches followed the prototype; the first was a batch of five, Marine Numbers 991–995, followed by a batch of three, Marine Numbers 1211–1213, followed by a final batch of eight, Marine Numbers 1372–1379, for a total of 17 airplanes.

WD11 torpedo bombers were operational in the Baltic and North Sea, but they were reported to be underpowered – the Gotha G-types of similar size and weight had 260 hp Mercedes D.IVa engines – and too lightly built for the severe demands put on them.

Torpedo attacks proved very difficult and required ideal sea and weather conditions. A drag rod under the fuselage was lowered prior to the attack; it was a device to measure height over the water, and the pilot had to maintain an altitude of 5 meters for torpedo release. Optical height measuring devices were also tested. Because of the difficulty of torpedo attacks coupled with the vulnerability of the aircraft to ship-board anti-aircraft guns, torpedo attacks were soon abandoned and the WD11s were modified to carry ten 50-kg bombs or a single Teka anti-shipping mine.

Gotha WD11 Specifications

Engines:	2 x 160 hp Mercedes D.III	
Wing:	Span Upper	22.50 m
	Span Lower	21.00 m
	Area	104 m^2
General:	Length	13.45 m
	Height	4.75 m
	Empty Weight	2146 kg
	Loaded Weight	3583 kg
Maximum Speed:		120 km/h
Climb:	1000m	12 min
	1500m	20 min
Service Ceiling:		3200 m
Range:		500 km

Above: The prototype WD11, Marine Number 679, afloat in the Gotha factory pond. Unlike the earlier WD7, the WD11 had its engines mounted as pushers. The box radiators were mounted directly above the engines.

Above: Front view of the prototype WD11, Marine Number 679, afloat in the Gotha factory pond.

Below: Gotha WD11 Marine Number 991 of the first production batch showing the Whitehead G/125 torpedo (45 cm diameter and 753 kg weight) hung in the semi-enclosed fuselage cavity. The wings and cowlings remain to be installed. On 15 June, 1917, this aircraft, crewed by *Lt.* Löwe & Thomsen, torpedoed and sank the S.S. *Kankakee* in the Thames Estuary.

Above: Front view of a WD11 emphasizing its distinctive Gotha appearance. A form of aileron servo is mounted on the outer rear interplane struts.

Below: Side view of the prototype WD11, Marine Number 679, afloat in the Gotha factory pond. The attachment and mounting of the aileron servo is clearly seen.

Above: WD11 being maneuvered by crane. This WD11 has the hexagonal naval camouflage on upper and side surfaces; the standard naval camouflage pattern specified blue-gray sides without hexagons. Unfortunately, the photograph is too dark to read the Marine Number, but Marine #1376 (last digit unclear, could be #1375) with shooting-star marking is second from right in bottom photo, page 62. The variety of markings on these Lindau-based WD11s indicates the shooting-star is a personal marking.

Below: Gotha WD11 showing the Whitehead G/125 torpedo hung just below the semi-enclosed fuselage cavity.

Above: Side view of the prototype WD11, Marine Number 679, on a beaching dolley.

Below: Gotha WD11 Marine Number 1376 of the third production batch was ordered in February 1917 and was attached to the *I. Torpedoflugzeug Staffel* in Windau. The torpedoman had elaborate aiming gear and a flexible machine gun.

Above: Two WD11s in flight taken from another aircraft.

Right: Two Gotha WD11s in flight with an Albatros W4 at bottom left.

Below: WD11s from the last production batch at Windau.

Above: Closeup of a WD11showing the forward gunner's station and left engine.

Below: Gotha WD11s and other seaplanes at Windau after the Zerel raid of 8 October 1917.

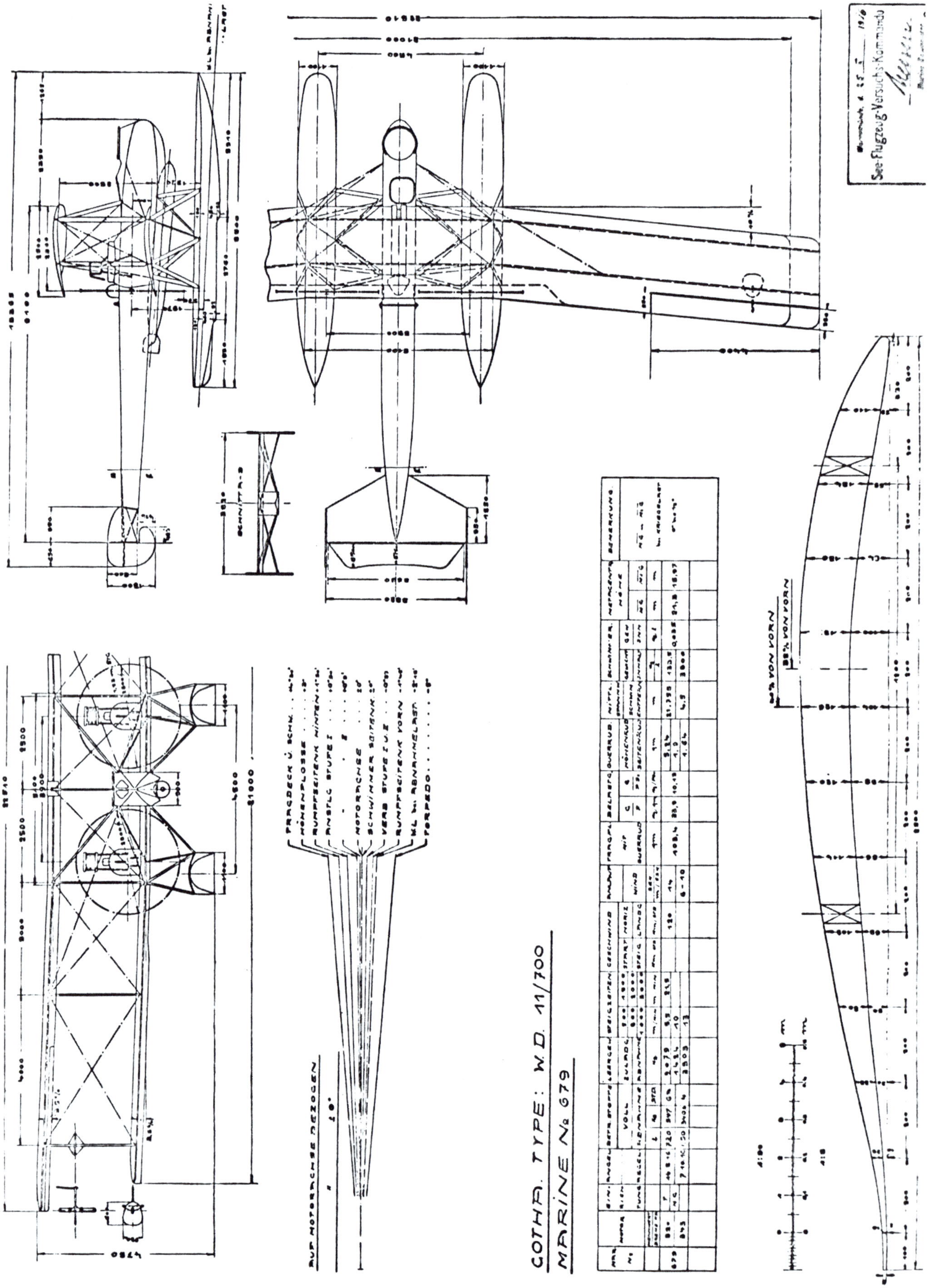
GOTHA TYPE: W.D. 11/700
MARINE No 679
See-Flugzeug-Versuchs-Kommando

Above: Previously identified as a WD14, this is a WD11 configured to carry bombs in place of a torpedo as shown by the pusher engine. The bombs were not carried internally but left to hang externally in this non-aerodynamic manner. The naval hexagonal camouflage was very distinctive on this WD11 and may indicate it had been recovered since being built.

Gotha WD12

The Gotha WD12 was designed by Rösner for maximum speed. Powered by a 160 hp Mercedes D.III engine, the WD12 was proposed by Gotha on 14 April 1916, and the prototype, Marine Number 944, was delivered to the *SVK* on 24 February 1917. Designated WD12a after modifications made in May to gain acceptance, the seaplane was fitted with an electrical bomb release for bombing trials. The WD12 was recommended for production because of its good seakeeping qualities. The WD12a was assigned to the flight school at Holtenau in February 1918.

On 28 August 1917 nine WD12s were ordered by Turkey in the Gotha work number batch 2150–2166, even numbers. Six had the 160 hp Mercedes D.III engine like the prototype; the other three had the 150 hp Benz Bz.III. The observer in the rear cockpit of these aircraft had a flexible Parabellum LMG 14 machine gun. These aircraft have been identified in some German Navy documents as WD9s, but the original Gotha factory photos are labeled WD12.

Gotha WD12 Specifications

Engine:	160 hp Mercedes D.III	
	150 hp Benz Bz.III	
Wing:	Span Upper	15.00 m
	Span Lower	14.00 m
	Area	54 m²
General:	Length	10.00 m
	Height	3.82 m
	Empty Weight	1000 kg
	Loaded Weight	1560 kg
Maximum Speed:		140 km/h
Climb:	1000m	7.5 min
	3000m	41 min
Service Ceiling:		4200 m
Range:		775 km

Above: The prototype WD12 showing the observer's faired headrest, consistent with this aircraft being unarmed.

Above: One of the Turkish WD12 seaplanes showing the aft gunner's cockpit with gun ring. This distinguishes the Turkish WD12s from the unarmed prototype used by the German Navy.

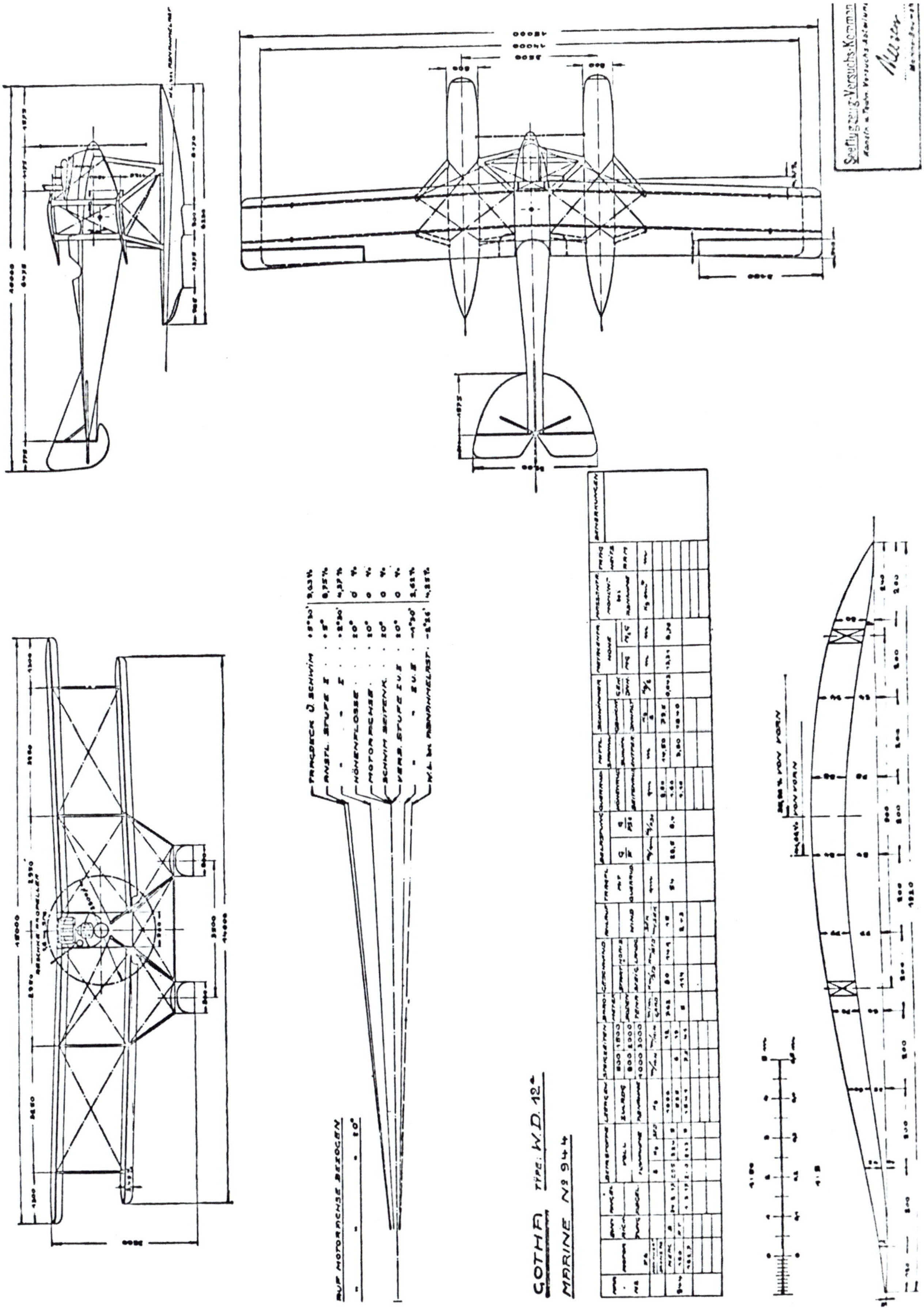
GOTHA TYPE: W.D. 12a
MARINE No 944
Seeflugzeug-Versuchs-Kommando

Gotha WD13

The Gotha WD13 was designed by Rösner and Hartwig. Powered by a 150 hp Benz Bz.III engine, the WD13 was not used by the German Navy; all six aircraft produced were to fill a Turkish order. These aircraft, Gotha works numbers 2186–2196, even numbers, were delivered between 21 December 1917 and 4 February 1918. The aircraft were accepted at the *SVK* and then shipped to Turkey. The observer in the rear cockpit of these aircraft had a flexible Parabellum LMG 14 machine gun.

Gotha WD13 Specifications		
Engine:	150 hp Benz Bz.III	
Wing:	Span Upper	14.60 m
	Area	49 m^2
General:	Length	10.07 m
	Height	3.74 m
	Empty Weight	1060 kg
	Loaded Weight	1460 kg
Maximum Speed:		131.5 km/h
Climb:	1000m	9 min
Duration:		3 hours

Above: Side view of a WD13 in Turkish markings on a beaching dolly. The observer's gun ring is clearly visible. Unlike most preceding Gotha seaplanes, the rudder has no aerodynamic balance. The German Navy evaluated the WD13 but did not purchase any; all six WD13s built were purchased and operated by Turkey.

Gotha WD Seaplanes Sent to Turkey

The table on page 93 lists at least 26 Gotha WD seaplanes known to have been delivered to Turkey, but the total number is thought to be 32. Data from the Turkish website in the bibliography is not entirely consistent with German sources; the reader may make up his/her own mind about details.

Above: Side view of a WD13 in Turkish markings. The observer's gun ring is clearly visible.

Below: Gotha WD13 in Turkish markings. The German Navy evaluated the WD13 but did not purchase any; all WD13s were purchased and operated by Turkey.

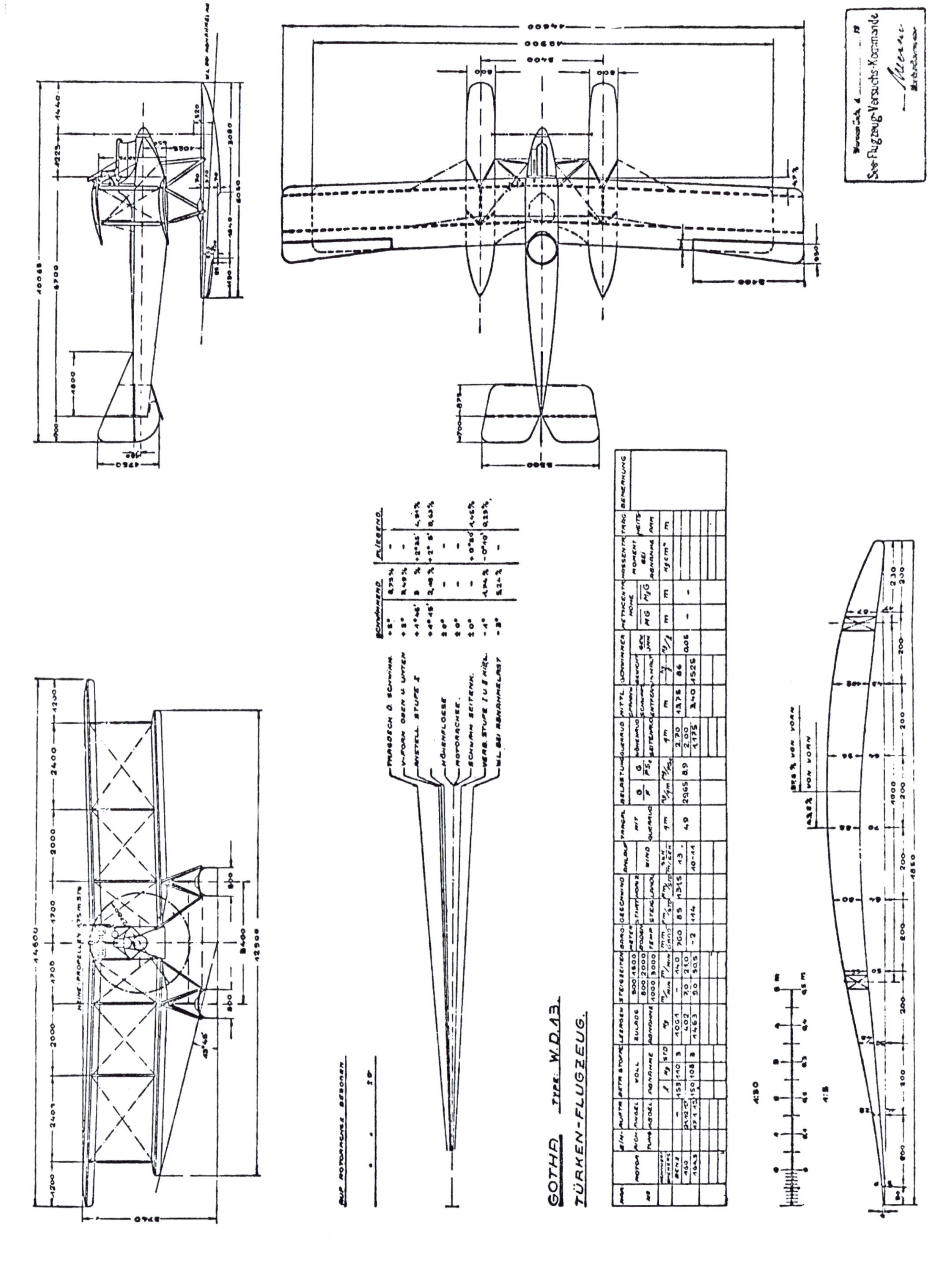
GOTHA TYPE: W.D.13
TÜRKEN-FLUGZEUG.
See-Flugzeug-Versuchs-Kommando

Gotha WD14

Designed by Rösner and Klaube, the Gotha WD14 was essentially an enlarged WD11 powered by more powerful 200 hp Benz Bz.IV engines mounted as tractors. The *SVK* praised the side-by-side seating of the pilot and observer, which facilitated coordination between them during target acquisition and torpedo launch. A third crewman had a flexible Parabellum LMG 14 machine gun and an aft turret.

During trials of the prototype, Marine Number 801, ailerons were added to the lower wings and the rudder area was increased to improve control with one engine out.

Gotha WD14 Specifications

Engines:	2 x 220 hp Benz Bz.IV	
Wing:	Span Upper	25.00 m
	Area	133 m^2
General:	Length	14.40 m
	Height	5.00 m
	Empty Weight	3090 kg
	Loaded Weight	5000 kg
Maximum Speed:		126 km/h
Climb:	1000m	17.1 min
	2000m	45 min
Service Ceiling:		3000 m
Duration:		1300 km

Above & Below: The prototype WD14, Marine Number 801, on a beaching dolly upon delivery to the *SVK* on 16 January 1917. This aircraft has a rear gunner but no nose turret; the pilot and observer sat side-by-side.

Above: The prototype WD14, Marine Number 801, upon delivery to the *SVK* on 16 January 1917. As a result of trials, ailerons were added to the lower wings and the rudder area was increased to improve control with one engine out.

Below: Delivered on 11 July 1917, WD14, Marine Number 1415 of the second series (Marine Numbers 1415–1430) had enlarged rudders and ailerons on all wings. Originally ordered as a torpedo bomber, it was modified for long-range maritime reconnaissance now that torpedo attacks had been basically abandoned. At Norderney, #1415 was used to test wireless equipment, navigation instruments, and droppable fuel tanks, here installed in the torpedo bay. Euler D.II 274/17, a single-seat trainer, provides an interesting size comparison.

Above: The prototype WD14, Marine Number 801, upon delivery to the SVK on 16 January 1917. As a result of trials, ailerons were added to the lower wings and the rudder area was increased to improve control with one engine out.

Below: Delivered on 11 July 1917, WD14, Marine Number 1415 of the second series (Marine Numbers 1415–1430) had enlarged rudders and ailerons on all wings. Originally ordered as a torpedo bomber, it was modified for long-range maritime reconnaissance now that torpedo attacks had been basically abandoned. At Norderney, #1415 was used to test wireless equipment, navigation instruments, and droppable fuel tanks, here installed in the torpedo bay. Euler D.II single-seat trainers flank it on both sides.

The first production series, Marine Numbers 1415–1430, had ailerons on all wings and balanced rudders. By the time these aircraft were delivered torpedo attacks had been shown to incur heavy losses with limited results, so the aircraft were modified for long-range reconnaissance with droppable fuel tanks in the torpedo bay. The next production series, Marine Numbers 1617–1631, had a gun turret in the nose and provision for bombs for use as a maritime reconnaissance bomber. The final production series, Marine Numbers 1946–1970, were designed for long-range maritime reconnaissance with droppable fuel tanks replacing the torpedo. This series had an additional central rudder for increased controllability, but were still viewed as unsuitable for poor aileron and rudder response. Only nine of this series were accepted before the Armistice.

A total of 69 WD14s were ordered (801, 1415–1430, 1617–1631, 1651–1662, 1946–1970), of which 52 were delivered (1629–1631 were cancelled and the Armistice stopped deliveries of the last 14). Designed as a torpedo bomber, the WD14 was re-cast as a long-range reconnaissance bomber after it was determined that torpedo attacks were ineffective and too dangerous. Moreover, the WD14 was underpowered and its control response was not satisfactory.

Above: WD14 Marine #1418 was wrecked while being flown by 1 *Torpedo-Staffel* pilot *Flgobmt.* Esser on 14 Jan. 1918.

Above: WD14 Marine Number 1617 of the third series was delivered in November 1917. To investigate its suitability as a reconnaissance bomber, it had a gun turret in a nose widened to hold five 10-kg bombs on each side. In place of a torpedo, ten 60-kg bombs could be carried in the torpedo bay. According to Navy records, twelve WD14s, Marine Numbers 1617–1628, were equipped for long-range reconnaissance flights of 8½ hours duration and twelve WD14s, Marine Numbers 1651–1662, were equipped for torpedo attack missions of 4 hours duration.

Above: WD14 Marine Number 1946 of the last series as photographed after modification on 14 June 1918. Equipped for long-range reconnaissance with a droppable fuel tank replacing the torpedo, the additional central rudder is just visible.

Above: This appears to be WD14 Marine Number 1946. This view clearly shows the additional central rudder added to the last series to improve control response. The new insignia was standardized 30 March 1918. WD14 #1946 was transferred from Warnemunde to Wiek or Bug on 2 October 1918.

Above: WD14 Marine Number 1429 was transferred from Flensburg to List on 17 December 1917, but was noted as being dismantled at List on 9 March 1918.

Above: WD14 Marine Number 1946 of the last series was photographed after modification on 14 June 1918. Twenty-five WD14s, Marine Numbers 1946–1970, were ordered in October 1917 equipped for long-range reconnaissance with a droppable fuel tank replacing the torpedo. For improved controllability this series had an additional central rudder. Reported "totally unsuited for the front" due to poor aileron and rudder response, the modified Marine Number 1946 is shown with balanced ailerons of increased area. Only nine WD14s of this series were accepted before the Armistice. The German naval camouflage scheme standardized in April 1917 specified three-color hexagonal camouflage on all upper surfaces, with side surfaces blue-gray and undersurfaces light gray. Despite that, photos show that production models of the Gotha WD14 wore the naval lozenge fabric over all surfaces except the floats, giving them a much darker appearance than the standard camouflage.

Below: WD14 cockpit with some instruments and controls labeled.

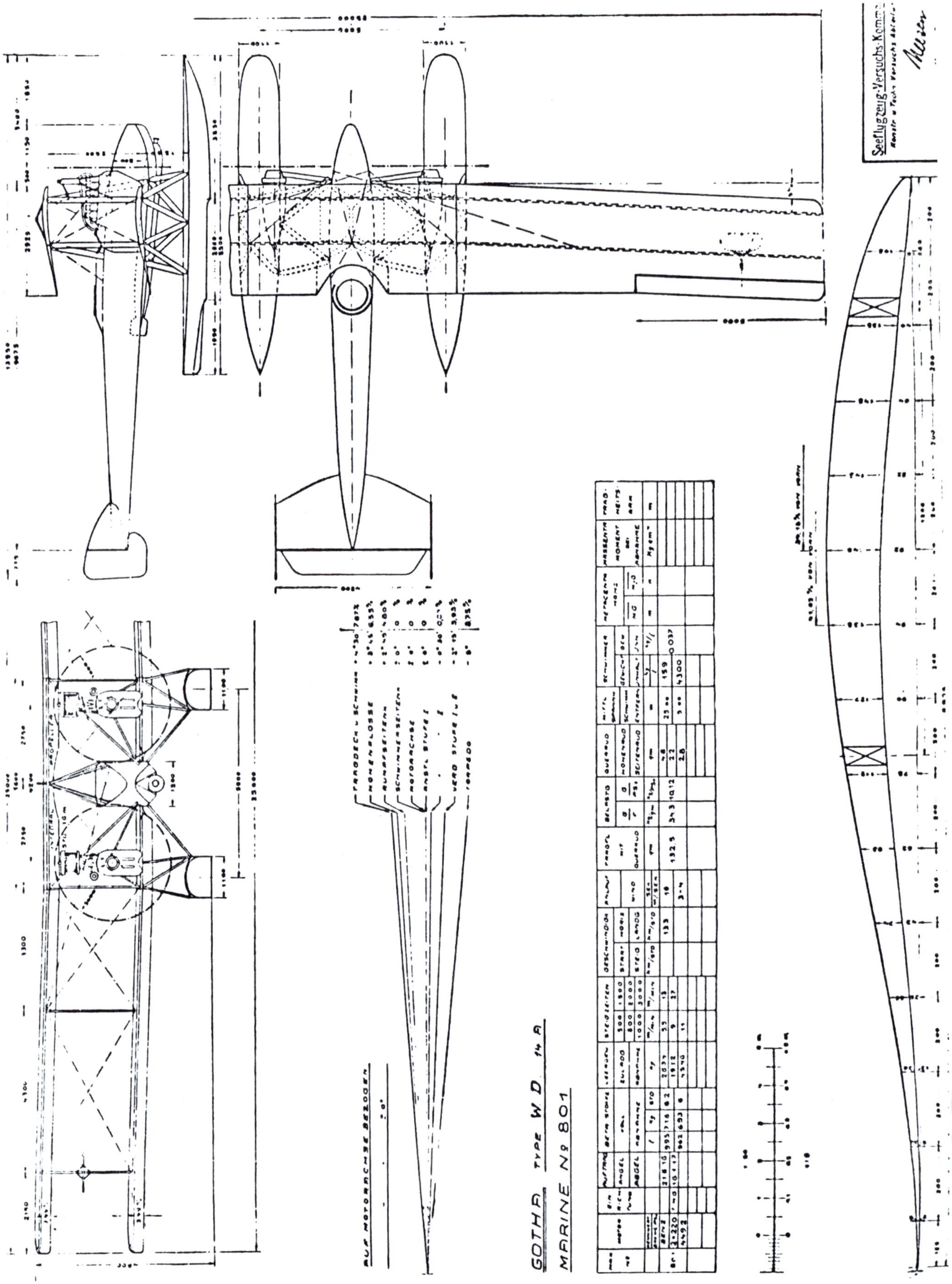
GOTHA TYPE W.D. 14 A
MARINE No 801
Seeflugzeug-Versuchs-Komm.

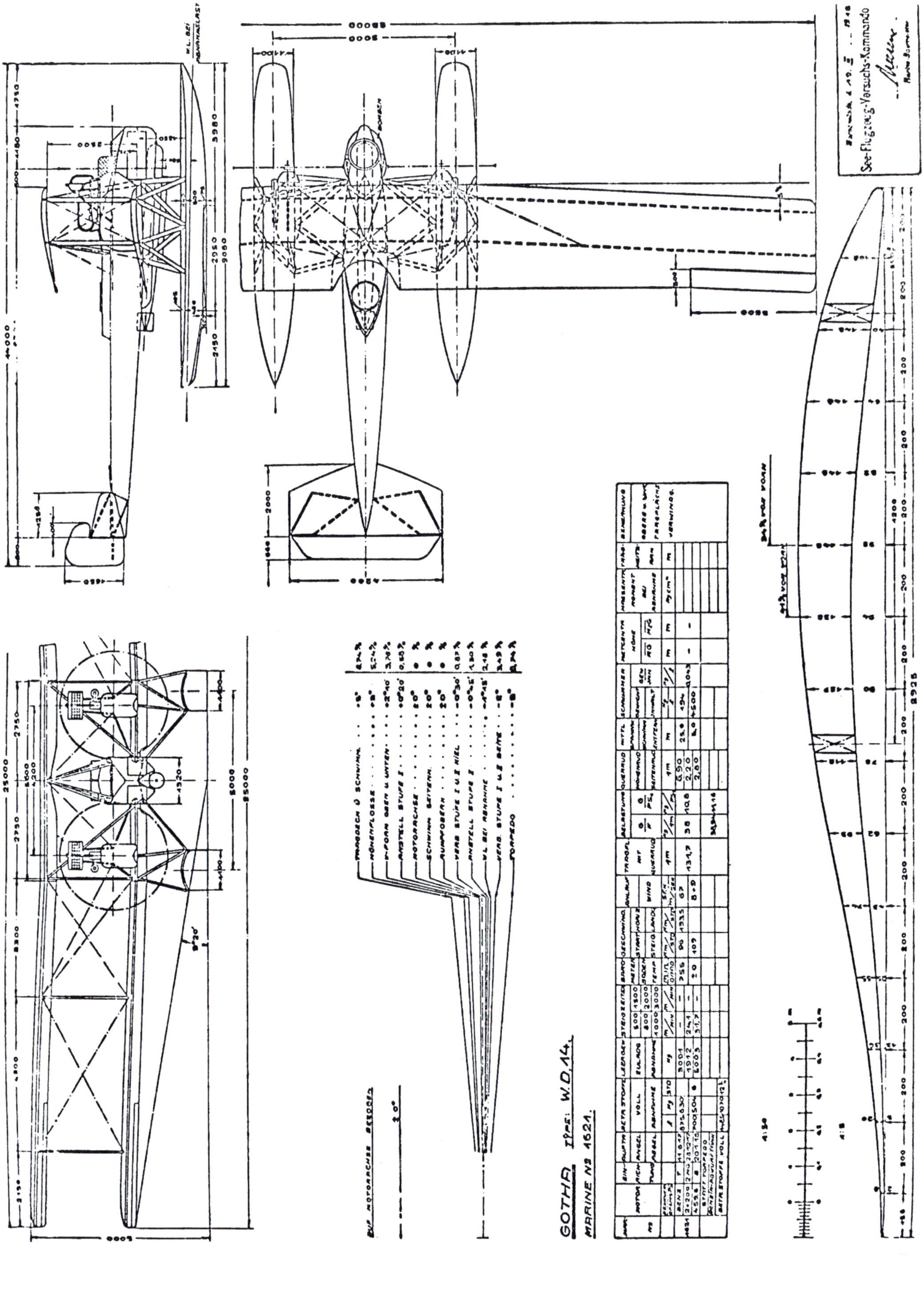
GOTHA Type: W.D.14.
MARINE No 1621.
See-Flugzeug-Versuchs-Kommando

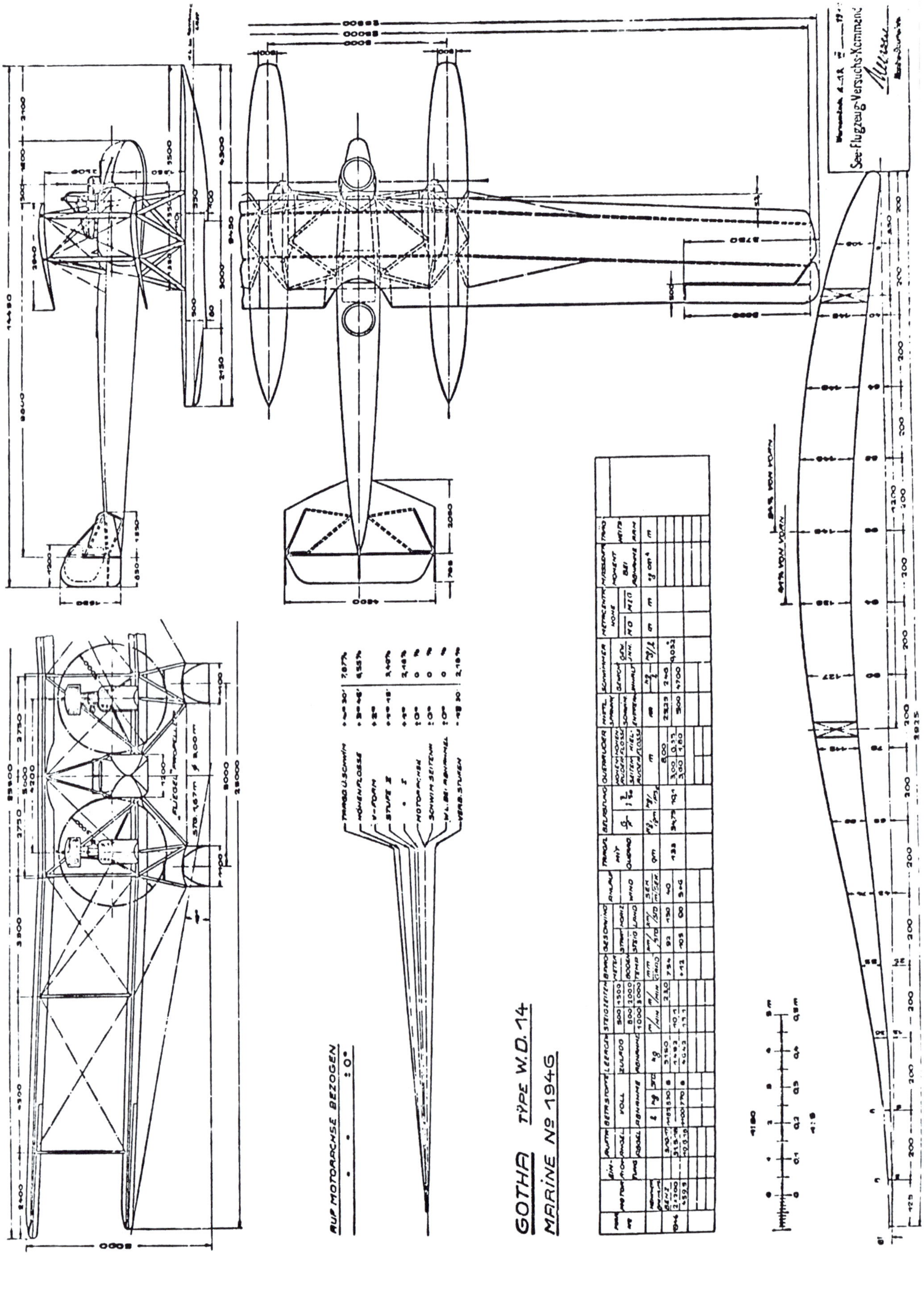
GOTHA TYPE W.D. 14
MARINE No 1946
AUF MOTORACHSE BEZOGEN
See-Flugzeug-Versuchs-Kommando

Gotha WD15

Designed by Rösner and Klaube, two Gotha WD15s were ordered on 14 July 1916 as heavy, multi-purpose seaplanes powered by a 260 hp Mercedes D.IVa engine. Although it had an excellent speed and rate of climb, the WD15 had engine problems despite being powered by a well-proven production engine.

The WD15 prototype, Marine Number 842, was destroyed on 18 September 1917 when a float collapsed. The second WD15, Marine Number 843, was delivered the next day and incorporated improvements as a result of flight tests of the prototype, including having ailerons on all wings with a strut connecting the upper and lower ailerons. The second WD15 was accepted in January 1918 but the *SVK* disapproved series production.

Gotha WD15 Specifications

Engine:	260 hp Mercedes D.IVa	
Wing:	Span Upper	17.20 m
	Span Lower	16.00 m
	Area	64.5 m^2
General:	Length	11.15 m
	Height	4.25 m
	Empty Weight	1544 kg
	Loaded Weight	2300 kg
Maximum Speed:		150 km/h
Climb:	1000m	8.5 min
	3000m	42.5 min
Service Ceiling:		4200 m
Duration:		900 km

Above: The first prototype WD15, Marine Number 842, on a beaching dolly upon delivery to the SVK.

Below: The first prototype WD15, Marine Number 842, in the Gotha factory pond before the Marine Number was added.

Above: The first prototype WD15, Marine Number 842, before the Marine Number was added. The first prototype had ailerons on the upper wings only.

Above: The first prototype WD15, Marine Number 842, on dollies. The 260 hp Mercedes D.IVa engine gave the WD15 good performance but both prototypes suffered from engine problems despite the Mercedes D.IVa being a generally reliable engine in mass production and use. The Gotha bombers used the Mercedes D.IVa without significant problems, so the numerous problems with the D.IVa in the WD15 were unusual.

Above: The first prototype WD15, Marine Number 842, on beaching dollies.

Above: The first prototype WD15, Marine Number 842, on beaching dollies. The typical Gotha rudder with horn balance underneath is evident.

Above: The second prototype WD15, Marine Number 843, afloat in the Gotha factory pond. The second prototype WD15 incorporated a number of improvements as a result of flight testing the first prototype; the most visible was the use of ailerons on all wings, with the upper and lower ailerons connected by an actuating strut. Like the first prototype, the second prototype also suffered a number of engine problems with its 260 hp Mercedes D.IVa engine, and despite being accepted in January 1918, the *SVK* did not approve series production.

German Naval Aviation Unit Designations

SFA I was neither a flying unit nor was it at Seebrügge. This designation, often mistaken for a flying unit, actually was a higher command organization. The I.*Seeflieger Abteilung*, or I.*SFA*, was the higher command responsible for managing all Eastern Front seaplane units, and was located at Kiel-Holtenau. Its equivalent, II.*SFA* at Wilhelmshaven, managed all seaplane units on the Western Front. The Germans never used "*SFA*" in any other context than in referring to one of these two regional commands.

The seaplane station at Seebrügge was actually known as *Seeflugstation Flandern* I, or sometimes *Flandern* I. Initially it was just *Seeflugstation Flandern*, but after the establishment of Ostend, it was referred to as *Seeflugstation Flandern* I, or sometimes locally as *See* I, while Ostend was *Seeflugstation Flandern* II (sometimes *See* II). *See* I has sometimes been confused with I.*SFA*.

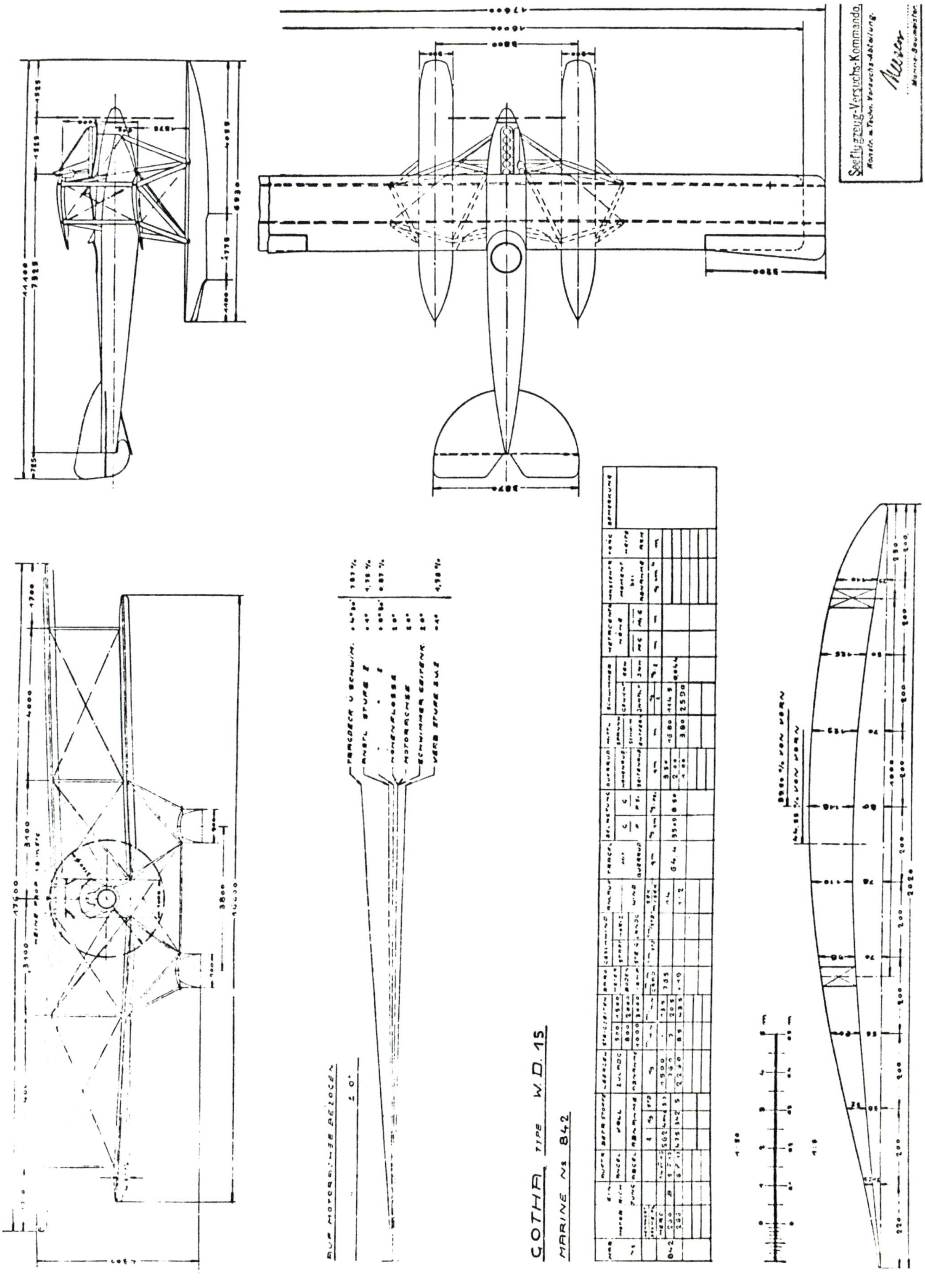
Seeflugzeug-Versuchs-Kommando
GOTHA TYPE W.D.15
MARINE Nr 842

Gotha WD20

The Gotha WD16–19 designs were not built; the next Gotha seaplane design to appear was the WD20. Designed by Rösner and Klaube, three Gotha WD20s Marine Numbers 1515–1517, were ordered in April 1917. Powered by two 260 hp Mercedes D.IVa engines, the WD20s were designed to carry the larger 50cm torpedo. However, experience soon revealed the limitations of torpedo attacks and the WD20s were reconfigured for long-range maritime reconnaissance in the North Sea. Despite removing three fuel tanks, the wireless operator and equipment, and two machine guns, and incorporating the maximum sweepback allowable, the WD20 was considered unacceptable for operational service due to poor flying characteristics. This was similar to the problems plaguing the G.IV and G.V bombers, indicating a fundamental problem with Gotha designs of the period.

Gotha WD20 Specifications

Engines:	2 x 260 hp Mercedes D.IVa	
Wing:	Span Upper	25.50 m
	Area	131.7 m^2
General:	Length	14.40 m
	Height	5.00 m
	Empty Weight	3030 kg
	Loaded Weight	4540 kg
Maximum Speed:		126 km/h
Climb:	1000m	15 min
Service Ceiling:		2000 m
Duration:		700 km

Above: Gotha WD 20 Marine Number 1515 in the factory before delivery in May 1918. Power was two 260 hp Mercedes D.IVa engines. Gotha WD 20 Marine Number 1517 had an additional, central rudder similar to that fitted to late-production WD14s. The WD20 was not accepted for operational service due to poor flying characteristics due to being tail heavy, problems similar to those plaguing the contemporary Gotha G.IV and G.V bombers.

Gotha WD22

The Gotha WD21 was not built; the next Gotha seaplane design to appear was the WD22. Designed by Rösner and Klaube, two Gotha WD22 long-range maritime reconnaissance seaplanes, Marine Numbers 2133–2134, were ordered in October 1917 for comparison with the Navy's giant Staaken floatplanes. The first WD22, Marine Number 2133, had two tractor 160 hp Mercedes D.III engines and two 100 hp Mercedes D.I pusher engines mounted in tandem.

Delivered in May 1918, WD22 Marine Number 2133 was found to be seriously tail heavy, seemingly an endemic problem with Gotha twin-engine and multi-engine airplanes. It was returned to the factory for increased sweepback and was again delivered to the *SVK* in September 1918.

The second WD22, Marine Number 2134, was delivered on 21 August 1918. It featured reversed engine positions; the Mercedes D.III engines now serving as pushers and the D.I engines as tractors.

Both WD22s were found at the Hage storage depot in December 1918.

Gotha WD22 Specifications

Engines:	2 x 160 hp Mercedes D.III and 2 x 100 hp Mercedes D.I	
Wing:	Span	26.00 m
	Area	147 m^2
General:	Length	14.35 m
	Height	5.20 m
	Empty Weight	3830 kg
	Loaded Weight	5200 kg
Maximum Speed:		131 km/h
Climb:	1000m	15.2 min
Service Ceiling:		3000 m
Duration:		780 km

Below: The first of two prototype WD22 seaplanes, Marine Number 2133, in its original form.

Above: The first WD22, Marine Number 2133, photographed at the SVK on 3 July 1918 in its original form. The tractor engines were 160 hp Mercedes D.III engines and the pushers were 100 hp Mercedes D.I engines.

Above: The massive WD22 would seem to deserve four 260 hp Mercedes D.IVa engines of the type used in the Gotha G.IV and G.V instead of the smaller, less powerful engines fitted. That less powerful engines were installed was likely a result of the Royal Navy's Distant Blockade that was strangling German access to resources.

Above: The first WD22, Marine Number 2133, photographed at the *SVK*. Like the Gotha G.IV and G.V bombers and the preceding WD20 long-range reconnaissance seaplane, the WD22 was very tail heavy, a fault that seriously reduced longitudinal stability and therefore greatly compromised flight safety. There was no rationale for this in a seaplane.

Below: WD22 cockpit with some of the instruments and controls labeled; note the throw-over control wheel at right.

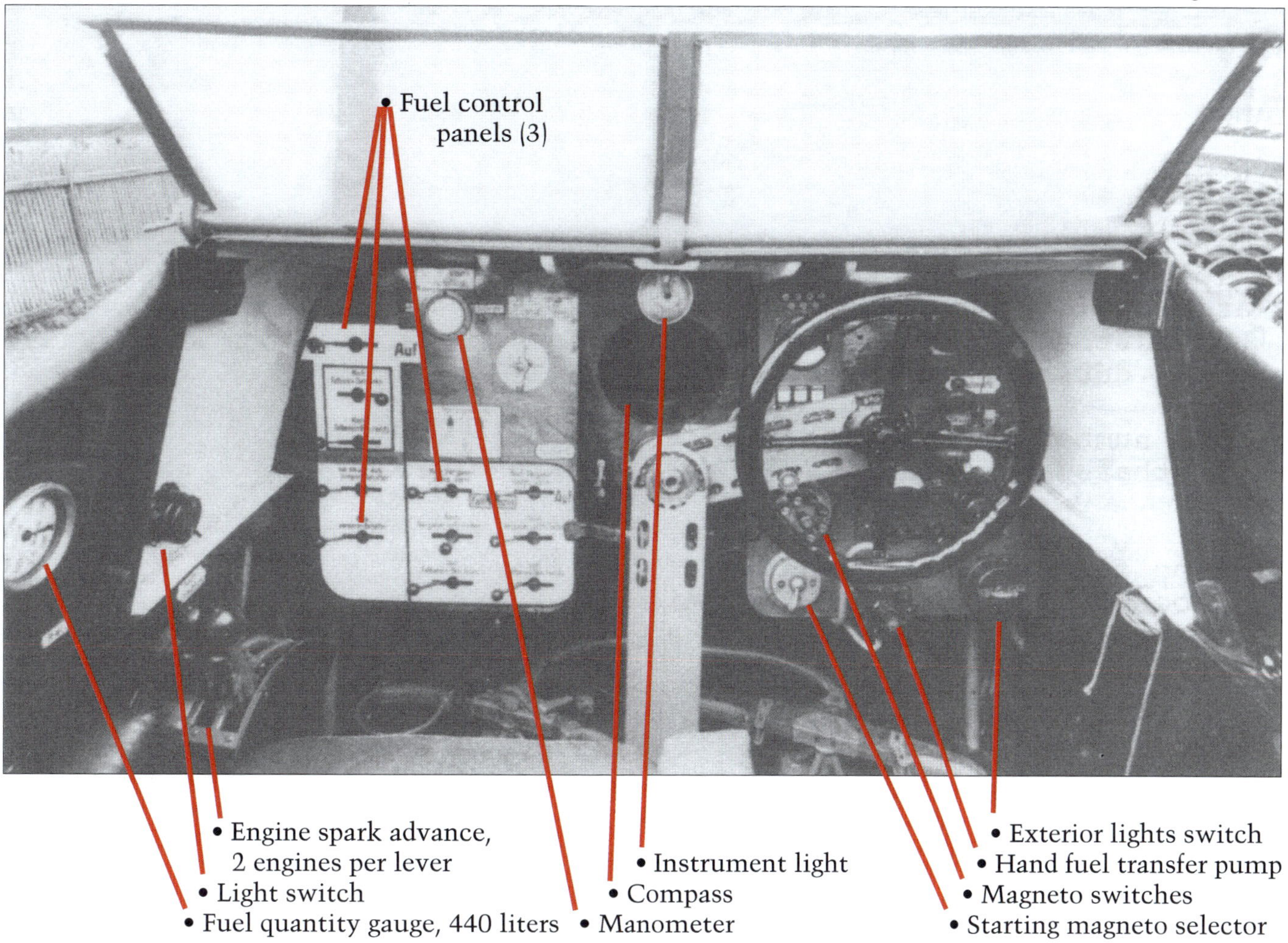

Gotha WD27

The Gotha WD23 through WD26 were not built; the final Gotha seaplane design to appear was the WD27. Designed by Rösner and Klaube, three Gotha WD22 long-range maritime reconnaissance seaplanes, Marine Numbers 4326–4328, were ordered for comparison with the Navy's giant Staaken floatplanes. The WD27s had four 175 hp Mercedes D.IIIa engines, two mounted as tractors and two in tandem mounted as pushers, enabling it to remain airborne with one engine out. This was difficult to achieve with the drag from a windmilling propeller; the fixed-pitch propellers of the time could not be feathered to reduce drag. The WD27 also featured folding wings to assist storage.

The WD27 was reported ready for factory engine trials in December 1918, unfortunately after the Armistice. Listed as complete in January 1919, the WD27 was later destroyed in accordance with Armistice requirements.

Gotha WD22 Specifications

Engines:	4 x 175 hp Mercedes D.IIIa	
Wing:	Span Upper	34.00 m
	Span Lower	31.00 m
	Area	197 m^2
General:	Length	17.60 m
	Height	6.00 m
	Empty Weight	4700 kg
	Loaded Weight	6860 kg
Maximum Speed:		135 km/h
Climb:	1000m	15 min
Duration:		1100 km

Above: The first WD27, Marine Number 4326, in the Gotha factory pond.

Below: Front view of the first WD27 on a beaching trolley.

Above: The first WD27, Marine Number 4326, on a beaching dolly. The duralumin floats were manufactured by the Zeppelin Werke Staaken.

Above: The first WD27, Marine Number 4326, on a beaching trolley. From right, the men are: *Ing.* A. Klaube, *Chefkonstrukteur Dipl. Ing.* K. Rösner, *Major a.D.E.* Blattmann, *Geh.Kom.Rat* A. Kandt, *Ob.Ing.* G. Rhein, *Ing.* F. Wagner. The rare Gotha G.X light bomber is at left.

Above: The first WD27, Marine Number 4326, in the Gotha factory pond. Great care was taken to streamline this huge aircraft by rounding the contours, smoothing the nacelles, and fairing the struts.

Gotha WD28

In July 1918 Gotha received a contract to build four WD28 seaplane prototypes, Marine Numbers 4001–4003 and 4034, based on the WD8 design. Powered by a 260 hp Mercedes D.IVa engine, the first two prototypes were configured as reconnaissance aircraft with wireless equipment and the last two were equipped to carry bombs. The first two prototypes were nearly complete when work was stopped in January 1919. No photographs appear to have survived.

Gotha Seaplane Production Summary			
Type	**Ordered**	**Marine Numbers**	**Remarks**
WD1	6	59, 285–289	286, 287, and 289 served in Turkey
WD2	22	60–61, 236–240, 254–258, 424–425	424–425 went to Turkey. Turkey ordered 8 more without Marine Numbers
WD3	1	259	Flown at Warnemünde as a trainer
WD4	1	120	Ursinus design, also known as UWD
WD5	1	118	118 was sent to Turkey
WD6	30	752–782	Brandenburg NW built under license, designated Brandenburg NW(Go)
WD7	8	119, 670–676	Torpedo trainers
WD8	1	476	Assigned to Zeebrügge, then sent to Turkey
WD9	1	572	Assigned to Zeebrügge, then sent to Turkey
WD10	1	782	Ursinus design, retractable float fighter
WD11	17	679. 991–995, 1211–1213, 1372–1379	Torpedo bomber, flew operational sorties
WD12	10	944	9 went to Turkey, no Marine Numbers
WD13	6	na	All 6 went to Turkey, no Marine Numbers
WD14	69	801, 1415–1430, 1617–1631, 1651–1662, 1946–1970	52 were produced; 1629–1631 were cancelled and 14 of the last batch were not delivered due to the end of the war
WD15	2	842–843	842 destroyed during testing
WD20	3	1515–1517	Not accepted for operational service
WD22	2	2133–2134	Too late for operational service
WD27	3	4326–4328	Too late for operational service. Only 1 built?
Notes: Quantity produced equals quantity ordered unless noted. Most seaplanes ordered by Turkey had no Marine Numbers assigned. At least 26 WD seaplanes went to Turkey to serve with Turkish units, and at least 32 are thought to have been sent; others served with German units in Turkey.			

Below: The Brandenburg NW was built under license by Gotha as the Brandenburg NW(Go). Gotha built 30 of these floatplanes that were given the internal Gotha company designation WD6.

SVK Records of Gotha Seaplane Orders & Production

Order Number	Type	Marine Numbers	Design	Class & Engine	1914												1915											
					J	F	M	A	M	J	J	A	S	O	N	D	J	F	M	A	M	J	J	A	S	O	N	D
1		59/61	WD1/2	B 150B						(3)	1	1				1												
2		118	WD5	B 150M													(1)								1			
3	672	119	WD7	C 2x120M													(1)											
4	120	120	UWD	C 2x160M													(1)											
5		236/240	WD2	B 160M														(5)					3		1			1
6		254/259	WD2	B 150B								(6)					1	2										1
7		285/289	WD1	B 100M													(5)	2	2									
8	476	476	WD8	C 240M															(1)									
9		572	WD9	C 160M																						(1)		
10		573/577	WD2	C.B. 150B																								(5)
11	672	670/676	WD7	C 2x100M																								
12	679	679	WD11	T 2x160M																								
13		752/781	Br.NW	B 160M																								
14	801	801	WD14	T 2x220B																								
15	842	842/843	WD15	B 260M																								
16	944	944	WD12	B 160M																								
17	679	991/995	WD11	T 2x160M																								
18	679	1211/1213	WD11	T 2x160M																								
19	679	1372/1379	WD11	T 2x160M																								
20	801	1372/1379	WD14	C2MG 2x200B																								
21	1515	1515/1517	WD20	T 2x260M																								
22	1621	1617/1628	WD14	T 2x200B																								
23		1629/1631	WD14	T 2x260M																								
24	1621	1651/1662	WD14	T 2x200B																								
25	1946	1946/1970	WD14	T 2x200B																								
26	2133	2133/2134	WD22	C3MG, 2x100M+2x160M																								
27		4326/4327	WD27	C2MGHFT, 4x160M, Long-Range Reconnaissance Floatplane																								
28		4328	WD27	C2MGHFT, 4x160M, Long-Range Reconnaissance Floatplane																								
Orders that Month										3		6						8	5	1						1		5
Deliveries that Month											1	1				1		1	4	4	1		3		2			2
Orders that Year																6												20
Deliveries that Year																3												17

Gotha Seaplane Orders and Production Notes

The table of Gotha seaplane orders and production above and on the facing page is from the *Atlas deutscher und ausländischer Seeflugzeuge* published by the *SVK* (*Seeflugzeug Versuchs Kommando* = Seaplane Testing Command), which was a classified document during the war. There is no official *SVK* data after June 1918 and further delivery information is fragmentary.

German Aerial Torpedo Operations in the North Sea

German aerial torpodo attacks in the North Sea started on 19 April 1917 and lasted until 9 September 1917. During this five-month period there were three successful torpedo attacks and a number of unsuccessful attacks. At least one of the successful attacks is credited to a Gotha WD11, Marine Number 991. See the adjacent table for a summary extracted from *The War in the Air*, Volume IV and *British Vessels Lost at Sea, 1914–1918* with supplementary notes from online discussion.

Aerial torpedo attacks were abandoned despite the fact the final attack was successful. This final attack was a joint action by torpedo planes and floatplanes armed with bombs. The German Navy decided the modest results did not justify the great effort required. Furthermore, ideal weather and sea conditions were required and the torpedo bombers were very vulnerable to ship-board defenses. Torpedo bombers were then reassigned as long-range reconnaissance seaplanes.

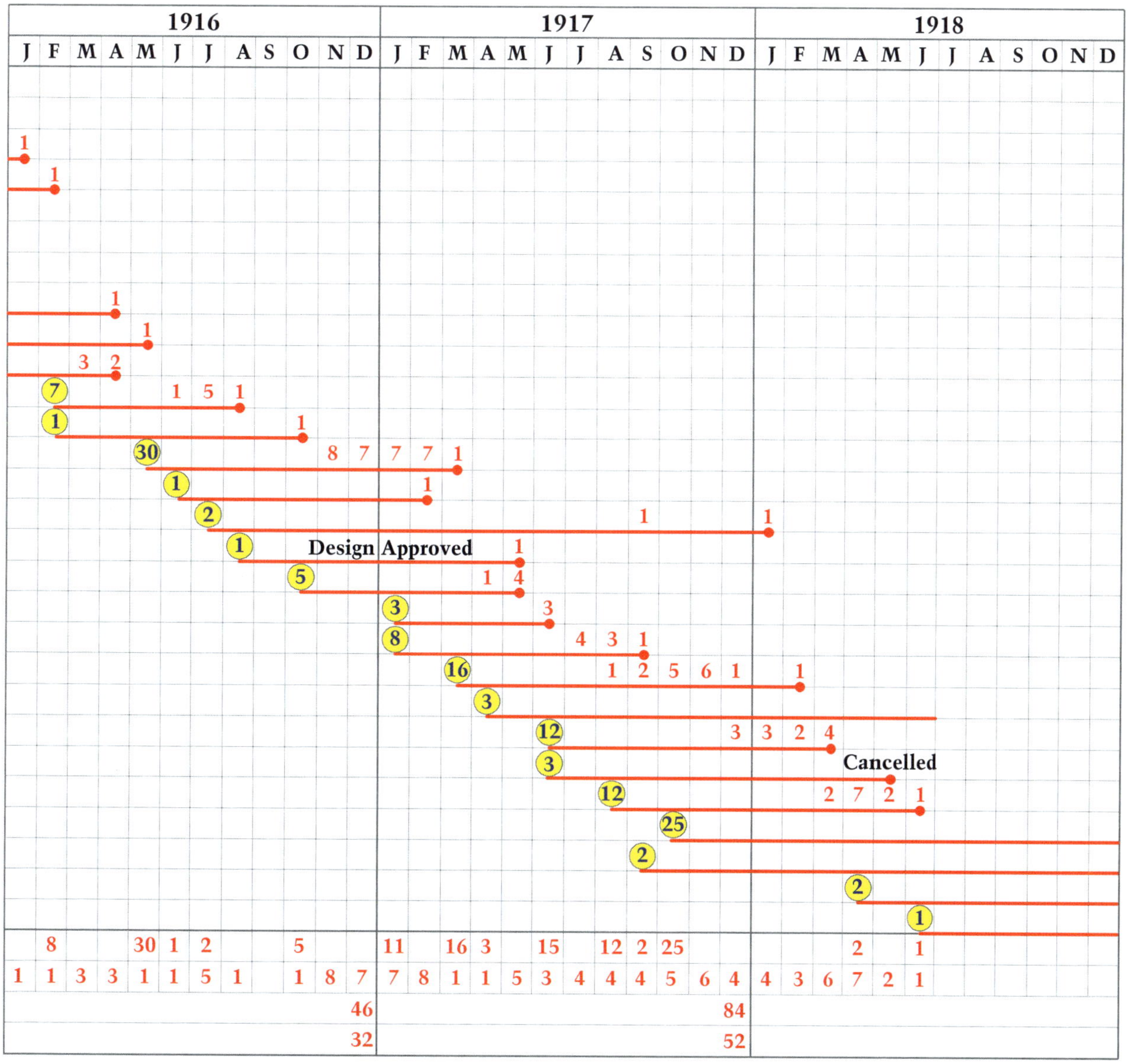

Summary of German Aerial Torpedo Attacks in the North Sea		
Date (1917)	**Ship**	**Remarks**
19 April	Drifter *Carolbank*	Torpedo missed
19 April	S.S. *Nyanza*	Torpedo missed
19 April	Monitor *Marshal Ney*	Torpedo missed
1 May	S.S. *Gena*	**Sunk** while underway (by *Lt.* Freude & *Flugmaat* Berghoff, 2.*Torpedostaffel* Zeebrügge?), but shot down escorting seaplane
20 May	Collier *Birchgrove*	Two torpedoes missed
14 June	Two unknown ships	Two torpedoes missed
15 June	S.S. *Kankakee*	**Sunk** (by WD11 torpedo plane T991, *Lt.* Löwe & Thomsen)
9 July	*Jarrix, Haslingden, Battersea, Hartley*	Five torpedoes missed
9 September	S.S. *Storm*	**Sunk**; hit by two of three torpedoes launched and hit by bombs

Gotha G Types

Gotha is best known for its series of G-types, or twin-engine bombers. Although these were used initially as short-range tactical bombers, the Gotha G.IV had sufficient range to permit bombing Britain from forward air bases along the Belgian coast. It was these bombing raids, and especially the daylight raids on London that started on 13 June 1917, that made Gotha a household name, and one to be feared.

Although damage from these raids was relatively minor, public outrage forced the British government to substantially build up the defenses of London; fighter squadrons were relocated to London from the Western Front and the anti-aircraft gun batteries around London were strengthened. The heavy guns assigned to London could not be used on the Western Front nor installed on ships to protect against surfaced U-boats.

In a matter of a few months the daylight anti-aircraft defenses of London were strong enough that the Gothas were forced to turn to night bombing operations over Britain, where they were soon joined by the newer Gotha G.V as well as Giant Staaken bombers. The night attacks were also of limited military value, but the need for Britain to mount a strong defense wasted resources that would have been better employed on the Western Front.

Bombs were normally loaded well forward of the Gotha bombers' center of gravity, but once the bombs were dropped the Gothas became tail heavy and pitch stability greatly deteriorated, making them difficult to fly and causing crashes on landing, normally the most dangerous part of flying. Landing crashes caused 76% of all Gotha bomber losses, more than three times losses from all other causes combined.

After the last bombing raid on Britain the night of 19/20 May 1918, the Gotha bombers returned to short-range tactical night bombing.

Above: Gotha G.IV of *Kagohl 3*, the *England Geschwader*, the unit assigned to bomb Britain. There are bombs under the extreme nose of the airplane; this unusual placement was due to the tail-heaviness of the Gotha bombers. Once the bombs were released the Gothas became neutrally stable or unstable in pitch, resulting in numerous landing accidents that destroyed more Gotha bombers than all other causes combined. At this remove it is difficult to understand why Gotha did not fix the tail-heaviness problem before the bombers were produced in quantity. Regardless, these bombers were Gotha's major success as an aircraft manufacturer. Without them Gotha would have little reputation in aviation.

Gotha G.I

At the dawn of aviation, the military authorities of the major powers considered how aviation might be used in war. Reconnaissance and bombing immediately suggested themselves as potential roles for aviation, and in fact both were important in WWI and since. Of course, the next thought was, how to defend against the enemy using his aviation assets against your own forces? This concern lead to anti-aircraft guns and the idea of using airplanes for air-to-air combat.

But what kind of airplane would be best for defeating other aircraft? In retrospect the answer is clear, but it was far from obvious before the war. Synchronizers to enable machine guns to fire between the blades of a rotating propeller were being designed, but got little notice at the time. The concept that resonated with many designers and military authorities before the war, before there was any experience in air-to-air combat, was the idea of aerial cruisers. The aerial cruiser theory was derived from the warships of the day; the pilot would fly the aerial cruiser within range of the enemy aircraft and gunners would fire flexibly-mounted machine guns and cannon to destroy the adversary. Thus was born the idea of the battleplane, and Britain, France, and Germany all built aircraft to this concept.

Friedel-Ursinus B.1092/14 Specifications

Engines:	2 x 100 hp Mercedes D.I	
Wing:	Span Upper	22.00 m
	Span Lower	19.00 m
	Chord (Upper & Lower)	2.2 m
	Sweepback	4°
General:	Length	17.60 m
	Height	6.00 m
	Empty Weight	4700 kg
	Loaded Weight	6860 kg
Maximum Speed:		90–95 km/h

Gotha G.I Specifications

Engines:	2 x 150 hp Benz Bz.III	
Wing:	Span Upper	20.30 m
	Span Lower	19.70 m
	Chord (Upper & Lower)	2.20 m
	Gap	1.95 m
	Sweepback	10°
	Area	82 m^2
General:	Length	12 m
	Height	3.9 m
	Empty Weight	1800 kg
	Loaded Weight	2966 kg
Maximum Speed:		130 km/h
Climb:	2000m	47 min

Above: The modified Friedel-Ursinus B.1092/14 and a captured Morane monoplane at *FEA* 9 in Darmstadt for a size comparison. Aerodynamically-balanced ailerons have been fitted and the nose radiators have been replaced by larger, twin radiators on each engine. A long cellon window is in the fuselage side over the wing.

Above: Three Gotha G.I aircraft stand on the Gotha airfield ready for flight testing. These may have been the first three built, G.9/15, G.10/15, and G.11/15, all of which were sent to *FEA* 7 to defend the Krupp steel works in Cologne.

Below: Closeup of the first Gotha G.I in the lineup above. The national insignia were painted below the top wing in addition to the usual locations. Both photos were released as Sanke cards.

Above: Gotha G.I 13/15 was delivered to *FEA* 3 on 2 September 1915, then flown to the Eastern Front. Enroute it landed at Schneidemühl where this photograph was taken. A streamlined bomb container is just visible between the wheels.

In early 1914 key German aviation authorities, including *Idflieg*, the *VPK* (*Verkehrstechnische Prüfungs Kommission* = transport technical investigation commission) and aviation industry executives, discussed the role of military aircraft and reached a consensus. The VPK then issued a directive outlining the tactical role of aircraft and specified three categories: *Typ I* was a fast two-seater intended for extended flights for reconnaissance and light bombing; *Typ II* was a light, maneuverable two-seater for short flights over the lines and armed for self-defense; and *Typ III* was three-seater designed to carry a large payload and fly low within range of enemy fire. The *Typ III* was required to have a speed over 120 km/h, climb to 800m in 10 minutes, have a flight duration of 6 hours, and a useful load of 450 kg. In essence the *Typ III* was a battleplane (*Kampfflugzeug* in German).

The German army approved the *VPK* recommendations on 28 April 1914, and demanded that these airplanes be developed as soon as possible. The *Typ I* was essentially the B-type two-seat biplane, of which numerous designs were available, while the *Typ II* eventually converged with the developed B-type to become the C-type armed two-seater. In parallel, a number of companies responded with designs to the *Typ III* requirements.

However, the design that was to evolve into the Gotha G.I stemmed from an independent effort. Oskar Ursinus, founder and editor of *Flugsport* magazine and a civil engineer, received orders to report to *FEA* 3 (*FEA* for *Flieger Ersatz Abteilung* = aviation replacement unit) in Darmstadt on 1 August 1914. On 9 August Ursinus proposed building a twin-engine *Kampfflugzeug* to *Major* Friedel, *FEA* 3's new commander, using under-utilized military personnel from the unit. Friedel accepted and the new airplane would be known as the Friedel-Ursinus *Kampfflugzeug*, also known as the type FU. The aircraft was certainly built with the *Typ III* requirements in mind; for example the fuselage and engine nacelles were armored.

Design work started immediately, and on 1 September *FEA* 3 personnel began to build the aircraft, which was assigned the military designation B.1092/14. This indicates *Idflieg* approved the aircraft and may have provided funds. On 30 January 1915 pilot Herold performed the aircraft's first flight.

The aircraft was powered by two 100-hp Mercedes D.I engines. The high fuselage enabled the engines to be very close to the centerline, reducing asymmetric control forces in event of engine failure. It also gave a wide field of fire to the gunner in the nose. However, with no protective structure above the crew, a turnover on landing would be extremely dangerous to the crew.

According to Ursinus's biography, the type FU was eventually sent to Ujatz, near Lodz, for operational

Above: Gotha G.I 42/15, named *Feodora*, was delivered to *FEA* 3 in Autumn 1915 before being sent to the Eastern Front.
Below: Gotha G.I 43/15 of the second production series.

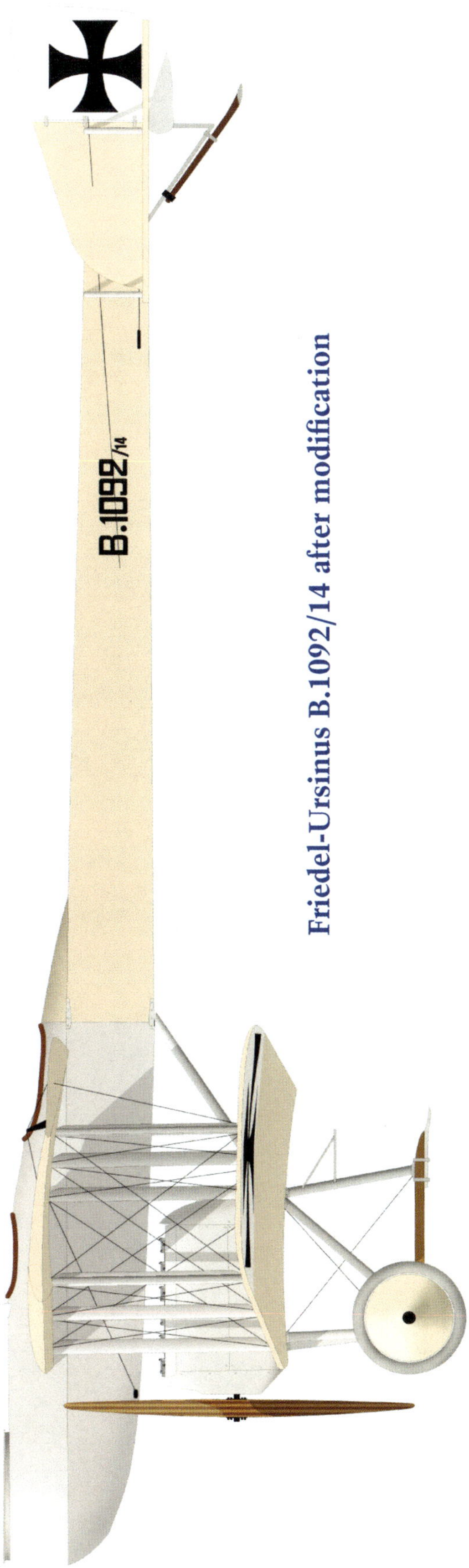

Friedel-Ursinus B.1092/14 after modification

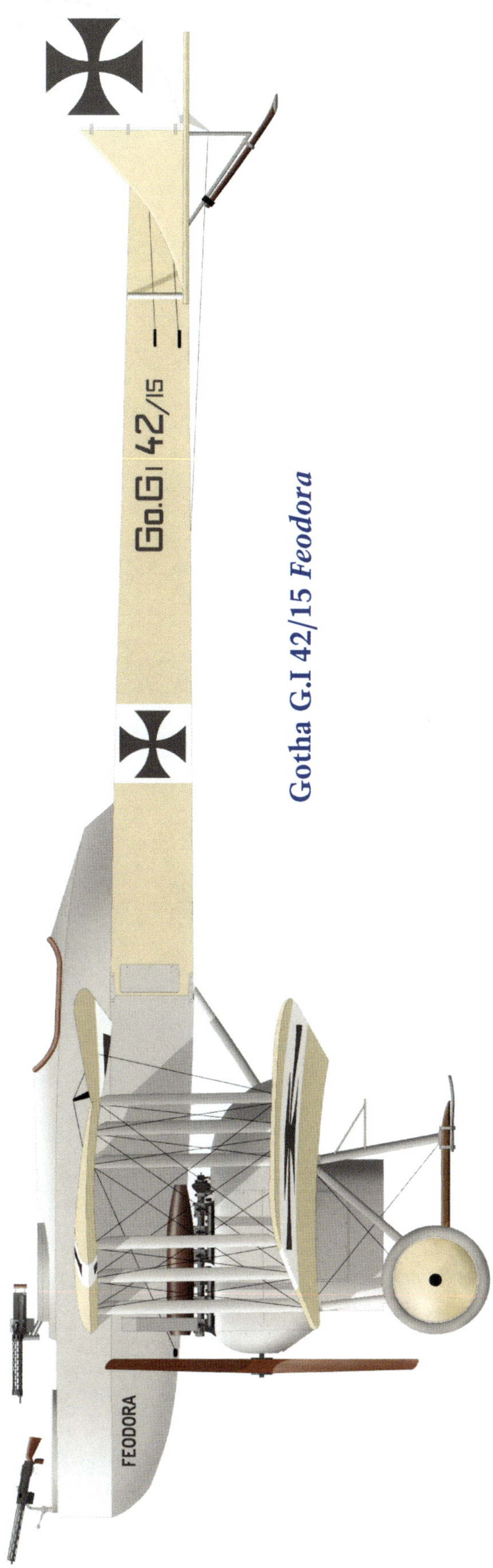

Gotha G.I 42/15 *Feodora*

Above: A Gotha G.I displays its distinctive silhouette in flight.

Above: This may be Gotha G.I G.41/15 of the second production batch, but the serial number is partly obscured.

Above: Engines running, a Gotha G.I is ready for take-off.

trials on the Russian Front, but there is no further information.

A production license for the type was offered to Fokker and Gotha; Gotha signed a license in March 1915. Likely in anticipation of that event, in February 1915 *FEA* 3 was transferred to Gotha. On 1 April 1915 *Idflieg* awarded a contract for six Gotha G.I aircraft to Gotha. The internal Gotha company designation was type UKL or type GUK; both were used. Of the first batch of six aircraft, five were to be powered by the 150 hp Benz Bz.III and the other was to be powered by two 160 hp Mercedes D.III engines. The contract required a crew of two with one machine gun, 200 kg of bombs, 150 kg of armor, and a maximum speed of 125 km/h.

Ursinus worked with Gotha engineer Hans Burkhardt to prepare manufacturing drawings. Starting on 27 July the aircraft were delivered as G.9/15–G.14/15; the last was delivered on 8 September. The first three went to *FEA* 7 for defense of the Krupp works in Cologne, the next two went to *FEA* 3 at Gotha for training, and the last went to *Armeeabteilung* Falkenhausen. The production G.I differed in detail from the Type FU.

On 15 July 1915 a second series of six Gotha G.I aircraft were ordered, all to be powered by the 150 hp Benz Bz.III. These aircraft were accepted between 22 September and 5 November, and all remained at *FEA* 3 in Gotha.

The third and final series of six Gotha G.I aircraft was ordered on 17 October 1915. These were powered by the 160 hp Mercedes D.III engine. The *Kampfflugzeug* concept now having failed, bombing was emphasized and the required bomb load was raised to 350 kg. A third crewman with a second gun carried between the pilot and front gunner was now specified. *Idflieg* also asked that a machine cannon be installed in addition to a machine gun, leading to weapons trials with a 20mm Becker cannon and 37mm cannon. The final batch was delivered between 24 January 1916 and 20 March 1916. One machine went to *FEA* 1 at Döberitz; the other five went to the *Prüfanstalt und Werft* (*Idflieg's* test establishment and workshop) at Döberitz.

In service the slow G.I accomplished little as a battleplane, the basic concept being fundamentally flawed. Battleplanes of all designs soon demonstrated they were too slow to catch enemy aircraft.

Above: The battleplane heritage of the Gotha G.I is demonstrated here. Oskar Ursinus handles a 20mm Becker cannon in the nose turret while the second gunner demonstrates the Parabellum machine gun; the pilot is seated aft.

Furthermore, even when the enemy was within range, so were they; the enemy had as good a chance at victory as they did. Operational experience soon showed that large, twin-engine types were better suited for bombing than air superiority missions, leading to the last batch of G.Is being modified for a greater bomb load. More specifically to the unique design the G.I inherited from the Type FU, turnovers on landing were always a possibility and, if they occurred, were invariably fatal to the crew. Moreover, the G.I was structurally fragile.

The German Navy purchased one float-equipped version of the G.I as the Gotha UWD, Marine Number 120.

Gotha G.I Production Summary				
Production Batch	**Serial Numbers**	**Qty**	**Engines**	**Crew/Armament**
#1 (ordered 1-4-15)	G.9/15–G.14/15	6	2x150 hp Benz Bz.III (five) 2x160 hp Mercedes D.III (one)	2 crew, 1 gun, 250 kg bombs
#2 (ordered 15-7-15)	G.40/15–G.45/15	6	2x150 hp Benz Bz.III	2 crew, 1 gun, 250 kg bombs
#3 (ordered 17-10-15)	G.100/15–G.105/15	6	2x160 hp Mercedes D.III	3 crew, 2 guns, 350 kg bombs
The Friedel-Ursinus prototype was assigned designation B.1092/14. It had one gun.				

Facing Page: Oskar Ursinus demonstrates the use of the forward gun in a Gotha G.I. The streamlined bomb container between the wheels is cleary visible and the nose of a 20-kg Carbonit bomb just protrudes from the lower front. Together with the bomb-dropping chute beneath the gunner's cockpit the bomb container is a clear indication of the evolution of the battleplane into a bomber. The propellers rotate in opposite directions to minimize torque.

Gotha G.II

Construction of the Gotha G.I gave Gotha the experience needed to design and build a long-range bomber, something *Major* Wilhelm Siegert, the commander of *Idflieg* and a supporter of strategic bombing, had wanted since the beginning of the war. Gotha engineer Hans Burkhardt, who had worked with Oskar Ursinus to build the Gotha G.I, was the natural choice for designer of the new bomber.

Earlier Burkhardt had modified crashed Gotha G.I G.9/15 by placing the fuselage on the lower wing, which greatly reduced the likelihood of a nose-over on landing. Burkhardt stated he had three main priorities when he designed the Gotha G.II; speed, protection of the observer in the nose, and ease of transportation. The later sounds odd today, but at that time, when airplanes were transported long distances, they were normally dismantled and moved by train rather than flying them. This counter-intuitive procedure was driven by the limited reliability of contemporary aircraft.

Burkhardt's ideas were accepted by *Idflieg*, who placed a production order for ten bombers on 18 December 1915. To have a worthwhile range and payload the new bomber needed much more power than the Gotha G.I, and the engines were specified as 220 hp Mercedes D.IV straight-eights mounted as pushers. The Mercedes D.IV was an inline eight-cylinder motor developed from the reliable 160 hp Mercedes D.III six-cylinder by adding two more

Gotha G.II Specifications

Engines:	2 x 220 hp Mercedes D.IV	
Wing:	Span Upper	23.70 m
	Span Lower	21.90 m
	Area	89.5 m^2
	Chord Upper	2.30 m
	Chord Lower	2.30 m
	Gap	2.22 m
	Sweepback	1.5°
General:	Length	12.40 m
	Height	4.30 m
	Empty Weight	2182 kg
	Loaded Weight	3192 kg
Maximum Speed:		148 km/h
Climb:	3000m	28 min
	4000m	41 min
Range:		500 km

Above: The Gotha G.II prototype was a completely different aircraft than the G.I. Distinguishing characteristics include the 2-bay wing, 4-wheel undercarriage under each engine, and the too-small rudder with no fin.

Above: This side view of the Gotha G.II prototype gives a better view of the inadequate vertical tail surfaces. The control cables were let outside the rear fuselage, adding drag to the airframe.

cylinders. The Mercedes D.IV was powerful and reliable in single-engine airplanes, but sometimes suffered crankshaft failures in twin-engine airplanes due to flexing of the block. A transitional design, the engine was soon replaced in production by the simpler six-cylinder D.IVa.

The resulting Gotha G.II was a completely new design that established the configuration for

Above: At least five production Gotha G.II bombers are parked in front of the flight test hangar in which the VGO Giant aircraft were built. Nearest the camera is G.203/16 with two-blade propeller, then G.205/16 with four-blade propeller, with G.200/16 behind it. The greatly enlarged vertical tail with fixed fin and enlarged, three-bay wings of the production aircraft is clearly shown. The engine was the straight-eight 220 hp Mercedes D.IV.

Above: Gotha G.II 204/16 photographed at the Gotha factory. This was one of the aircraft delivered to *Kagohl* 4, *Staffel* 20 on 24 August 1916. The G.II airframe was the basis for the later G.III, G.IV, and G.V.

all subsequent operational Gotha bombers. The prototype G.II was a two-bay aircraft that entered flight trials in March 1916. The under-carriage had eight wheels, four underneath each nacelle, for safe landings. Both two-bladed and four-bladed propellers were tried with the slow-turning geared engine. Matching a fixed-pitch propeller with engine and airframe to maximize speed and climb was a painstaking task that could only be accomplished by time-consuming flight-testing.

The prototype G.II was fast for its time but had insufficient climb with a full bomb load. An enlarged, three-bay wing gave the production version the required performance.

The undercarriage was also revised for the production G.II. Although the original undercarriage prevented nose-overs, it lacked brakes as did all WW1 aircraft. An extended landing run could result in a fatal accident by running off the edges of the field. To solve this potential problem Burkhardt simplified the undercarriage by removing the forward wheels and moved the center of gravity aft, allowing a conventional tail skid that provided the necessary braking action. Unfortunately, this aft shift in the center of gravity made the aircraft far less stable in pitch, a situation that was to plague all subsequent Gotha bombers in service. Landing accidents due to pitch instability were responsible for 76% of all Gotha bomber losses, more than three times as many losses as all other reasons combined.

Finally, the too-small rudder of the G.II prototype was replaced with a fixed fin and larger rudder in the production G.II, improving stability and controllability in flight.

The G.II had a crew of three; a bombardier-gunner in the front cockpit, the pilot in the center cockpit, and a gunner in the aft cockpit. The wings and fuselage were of typical wood and fabric construction, although the nose was covered with plywood. The tail surfaces were of steel tubing covered with fabric. The engine nacelles and undercarriage formed a single sub-assembly; for easy handling these could be moved around when fitted with separate, auxiliary wheels, a feature that was pattented. Fuel and oil tanks were in the engine nacelles with a gravity fuel tank mounted above the upper wing. Two bomb racks in the fuselage held fourteen 10-kg bombs. The total useful load was 1,010 kg.

Production of the G.II began on 25 April 1916 and

Above: Gotha G.II 207/16; the dark nose is due to being covered in plywood; the rest of the airframe is covered in fabric.

the type test was completed on 17 July 1916. Serial numbers assigned to the ten production aircraft were G.II 200/16–209/16. The G.II wing cellule finally passed the static load test during the week of 11 August 1916 after six failures that had to be rectified.

Of the 10 G.II bombers built only eight reached *Staffel* 20 of *KG*4 in August/September 1916, apparently the only unit to use the G.II operationally. One G.II remained at *FEA* 3 in Gotha and the other was badly damaged during flight evaluation. Unfortunately, there is no information available regarding the combat use of the G.II, which was limited by the small number of aircraft and the problematical reliability of the engines.

Gotha G.III

The G.II was quickly superseded in production by the improved G.III. The main difference between the types was the engine; the G.III was powered by the 260 hp Mercedes D.IVa engine of six cylinders. The new engine did not suffer the occasional crankshaft failures of the straight-eight it replaced, was cheaper to produce, offered slightly more power, and was soon available in quantity. The additional power enabled the useful load to be increased from 1,010 kg to 1,235 kg.

The other important change for the G.III was an opening for a ventral, downward-firing machine gun. *Idflieg* ordered 25 G.III bombers on 3 May 1916; deliveries started on 16 October and were completed on 26 March 1917. Serial numbers are thought to be G.375/16–G.399/16. One G.III, G.398/16, was sent to *Halbgeschwader* I as a trainer and G.392/16 was sent to the Daimler factory for flight-testing an experimental Mercedes D.IVa. The rest of the G.IIIs were delivered to *KG*2. *KG*2 was moved around and flew missions on the Western Front as well as daylight bombing missions on the Balkan Front, stationed at Hudova along with a few G.II bombers. The most significant result in the Balkans was destruction of the railway bridge over the Donau at Cernovoda in late September 1916; this deprived the Romanian army of essential supplies and reinforcements. Interestingly, one squadron commander complained that the G.III easily out-distanced its two-seat escorts. In August–September 1917 the G.III was retired from the front.

Gotha G.III Specifications

Engines:	2 x 260 hp Mercedes D.IVa	
Wing:	Span Upper	23.70 m
	Span Lower	21.90 m
	Area	89.5 m^2
	Chord Upper	2.30 m
	Chord Lower	2.30 m
	Gap	2.22 m
	Sweepback	1.16°
General:	Length	12.22 m
	Height	3.90 m
	Empty Weight	2383 kg
	Loaded Weight	3618 kg
Maximum Speed:		140 km/h
Range:		700 km

Above: Gotha G.III 376/16 or 378/16; like the G.II, the dark nose is due to being covered in plywood; the rest of the airframe is covered in fabric.

Facing Page: Gotha G.III 398/16; differences between the G.II and G.III were slight and only visible from certain viewpoints.

Below: Gotha G.III; the bulge on the upper side of the cockpit provided room for the pilot's controls.

Above: This Gotha G.III wears light wheel covers with a dark stripe; the exact colors are not known.

Below: The Gotha G.III was used for tactical bombing in daylight; this one was downed on 8 February 1917 by famous French ace *Captaine* Georges Guynemer and *Adjudant* Chainat. Another was downed on 23 April 1917 by Sub-Lt. L.S. Breadner.

Above & Below: Two more views of the Gotha G.III downed on 8 February 1917 by French ace *Captaine* Georges Guynemer and *Adjudant* Chainat showing its markings as the wreckage was being re-assembled for display.

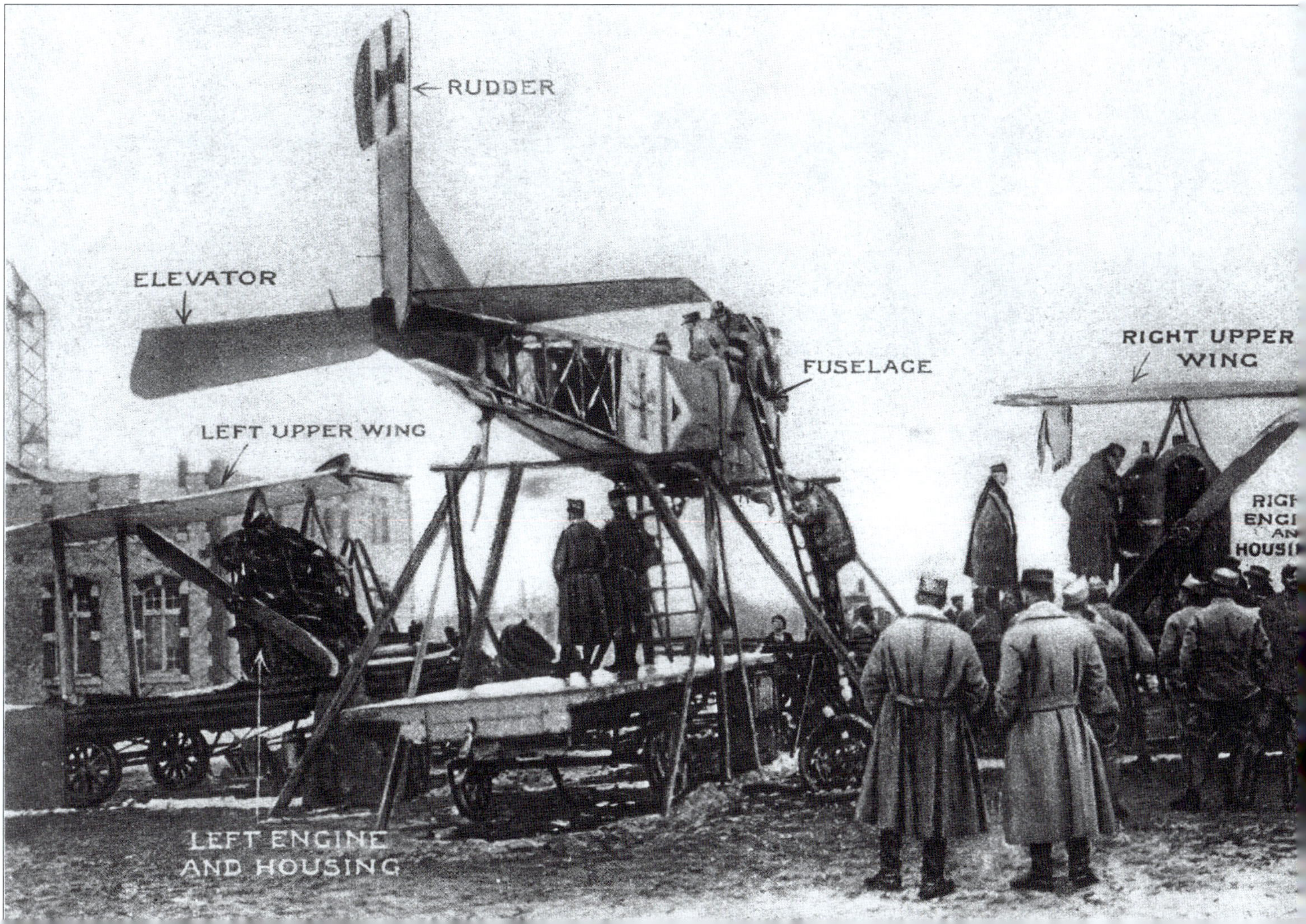

Gotha G.III of *KG2* downed 8 February 1917 by *Captaine* Georges Guynemer and *Adjudant* Chainat.

Gotha G.III of *KG2*, early 1917

Above & Below: This Gotha G.III 389/16 (or possibly 385/16) of *Kagohl* 2, *Staffel* 19 wears a macabre 'death's head' marking on the nose. The upper surface and sides of the fuselage and upper surfaces of the wings were painted green. The outer wheel covers were painted in the black and white halves typical of *Kagohl* 2. A black triangle with white outline is painted on the rear fuselage. Belts of signal flares are attached to the side of the bombardier's cockpit, and the bomb racks are visible under the wing center section.

Above & Below: Gotha, probably a G.III but possibly a G.II, on a peaceful flight displays the characteristic Gotha shape.

Gotha G.III of *KG2*, early 1917

Gotha G.III of *KG2*, early 1917

Gotha G.IV

By late 1916 it was apparent to the German Army that the strategic bombing campaign against Britain using Zeppelins was not a success, and furthermore the Zeppelins were too costly and vulnerable. The Army was planning to abandon airships in favor of cheaper, more effective heavy bombers. The operation to attack London and other strategic targets with bombers was called *Türkenkreuz* (Turk's Cross).

Of course, the plan required bombers with sufficient range and payload for the task, and the Gotha's performance made it the obvious choice. On 16 August 1916 *Idflieg* ordered 52 Gotha-built bombers of an improved type, the G.IV, making the G.IV the first Gotha design to be ordered in substantial numbers. So many of the new Gotha G.IV bombers were wanted that two other manufacturers, LVG and SSW, were given contracts to build the G.IV under license.

The G.IV was simply a refinement of the G.III. Powered by the same 260 hp Mercedes D.IVa engine used in the G.III, the G.IV featured ailerons on both upper and lower wings connected by an actuating strut for improved controllability. The G.IV also introduced the patented Gotha tunnel, a hollowed-out opening in the rear fuselage that enabled the gunner to depress his gun through the tunnel to fire at aircraft below, or to mount a ventral gun at floor level for a wider field of fire. To retain torsional rigidity despite the cut-out in the bottom of the fuselage, the fuselage was covered in plywood instead of fabric.

Gotha G.IV Specifications

Engines:	2 x 260 hp Mercedes D.IVa	
Wing:	Span Upper	23.70 m
	Span Lower	21.90 m
	Area	89.5 m^2
	Chord Upper	2.30 m
	Chord Lower	2.30 m
	Gap	2.22 m
	Sweepback	1.16°
General:	Length	12.40 m
	Height	4.30 m
	Empty Weight	2413 kg
	Loaded Weight	3648 kg
Maximum Speed:		140 km/h
Climb:	1000m	3 min
	2000m	9 min
	3000m	16.5 min
	4000m	25 min
Range:		700 km

Gotha G.IV(LVG)

LVG received an order for 150 Gotha G.IV(LVG) bombers in December 1916. The type test for the initial 50 aircraft was completed on 25 June 1917, after the Gotha-built aircraft had already bombed London. One aircraft was fitted with two 245 hp Maybach Mb.IVa engines in an attempt to raise the bombers' operational altitude, but inability to find a suitable propeller thwarted the attempt. Another

Below: Gotha G.IV 408/16 was a bomber assigned to *Kagohl* 3, the *England Geschwader*, the unit assigned to bomb Britain. Two fuel tanks are mounted above the upper wing to give it enough range for these missions.

Above: Gotha G.IV 408/16 of *Kagohl* 3 after full tactical markings were applied. The letters on the fuselage side are likely the initials of two of the crew members, a common marking practice in *Kagohl* 3.

aircraft tested Flettner servo tabs starting in October 1917.

LVG built an additional 40 G.IV bombers to an Austro-Hungarian order. These were modified to take 230 hp Hiero engines built in Austria. The squadrons received these aircraft in March–April 1918. Weak engine bearers and unsuitable propellers led to excessive engine vibration that caused piping leaks and structural damage, limiting their usefulness, and by September 1918 the G.IV(LVG) bombers were essentially grounded.

Gotha G.IV(SSW)

Idflieg ordered 80 Gotha G.IV(SSW) bombers from SSW at the same time the LVG machines were ordered. The first batch of 40 were delivered between July 1917 and February 1918; about 30 went to operational units and the rest went to training and replacement units. The second production batch, ordered in May 1917, was delivered to training units between December 1917 and August 1918. The twin-wheel *Stossfahrgestell* (shock landing gear) designed by Siemens was fitted to all aircraft from G.217/17, and many machines were also fitted with the Flettner servo tabs. Instead of the normal 260 hp Mercedes D.IVa engines, some aircraft had 185 hp NAG C.III or 180 hp Argus As.III engines for training use.

Gotha G.IV Operations

The G.IV was the aircraft that made Gotha a household name. Built in much greater numbers than earlier models – 232 were built by Gotha, LVG, and SSW – it was intended by the German Army to replace the expensive, vulnerable Zeppelins as Germany's long-range bomber of choice. Although the first G.IVs were delivered in November 1916, a number of problems had to be resolved before the G.IVs were ready to bomb London, their intended target, delaying their first attack on England to 25 May 1917. This attack by *Kagohl* 3 did not reach London, nor did the second. But on 13 June the third attack did reach London, causing the most severe casualties of any bombing raid of the war. Together with the heavy casualties, the spectacle of Gothas leisurely bombing targets in London in broad daylight shocked the British public, with a significant ramification being the creation of the RAF from the RFC and RNAS on 1 April 1918 to provide a more coordinated and effective air defense capability. In the interim, the war cabinet agreed to double the size of both the RFC and RNAS as a direct result of these raids.

Initial Gotha G.IV losses during the daylight attacks were light, but rapidly improving defenses caused a shift to night bombing starting the night of 3/4 September after eight daylight raids.

Above: Gotha G.IV 410/16 of *Kagohl* 3. The early Gotha G.IV bombers were painted light blue overall with natural metal engine cowlings.

The Gotha bombers were stable when fully loaded but only marginally stable when lightly loaded, as was normal during landing. Photos of loaded Gothas show bombs mounted under the front gunner's cockpit, far ahead of the center of gravity. Small, 12.5-kg bombs were also carried in bomb racks in the bombadier's cockpit. After release of these bombs the center of gravity moved aft significantly, making the Gothas much less stable in pitch. Accordingly, most operational losses were due to crashes during landing, when the lightly loaded bombers were least stable. And the night landings now required in its new role as night bomber emphasized the Gotha's handling problems during approach and landing; landing accidents causing 76% of all Gotha losses.

Below: Gotha G.IV 410/16 of *Kagohl* 3 after a crash during take-off. The "Gotha Tunnel" that enabled the rear gunner to fire downward and to the rear is shown to advantage.

Gotha G.IV 408/16 of *KG3*, May/June 1917

Gotha G.IV 604/16 of *KG3*, May/June 1917

Above: A factory photograph of Gotha G.IV 607/16 carrying six 50-kg PuW bombs beneath the fuselage. The far-forward carriage of the bombs under the nose illustrate the tail-heaviness of the basic design. The bombardier/forward gunner normally carried a number of 12.5 kg bombs in bomb racks inside his cockpit, which partially compensated for the center of gravity problems, at least until they were dropped.

Below: Gotha G.IV 602/16 crash-landed in neutral Holland at Sas van Gent on 28 September 1917 during a night raid on Britain. Another G.IV crashed and burned at Sneek the same day.

Above: A factory photograph of Gotha G.IV 601/16 after a landing accident. All aircraft design decisions are compromises between competing concerns, but landing accidents claimed 76% of the Gotha G.IV and G.V bombers lost to all causes. If the CG is too far forward, the aircraft is excessively stable, too much download is required on the tail, raising the nose to abort a landing becomes difficult, and nose-overs become more likely. However, the extreme number of landing accidents is proof that Burkhardt moved the center of gravity too far aft during development of the G.II, a mistake that could have been easily rectified but was not.

Below: An unidentified Gotha G.IV after yet another landing accident.

Gotha G.IV serial unconfirmed but thought to be 624/16, *Hptm.* Ernst Brandenburg, C.O. of *KG*3, May/June 1917. Brandenburg was awarded the *Pour le Mérite* for leading *KG*3 to bomb London

Gotha G.IV 405/16 of *Oblt*. Hans *Freiherr* Ulrich von Trotha, deputy commander of *KG*3, May/June 1917

Above: A Gotha G.IV camouflaged for night bombing missions in enemy hands.

Below: An unidentified Gotha G.IV at the factory. The twin-wheel *Stossfahrgestell* (shock landing gear) designed by Siemens was fitted to all SSW-built aircraft from G.217/17, and apparently this G.IV is one of those. Many SSW-built machines were also fitted with the Flettner servo tabs to reduce the control forces, making the aircraft more maneuverable and reducing the pilot's workload.

Above: General Hindenburg and party inspect a Gotha G.IV, probably 612/16 if the lettering on the wheel cover is correct. The extreme forward mounting of the bombs below the nose clearly indicates the aircraft was tail-heavy after bombs were released.

Below: Front view of an unidentified Gotha G.IV. From this vantage point the actuating strut between upper and lower ailerons confirms this is a G.IV and not a G.III, which did not have ailerons on the lower wings.

Gotha G.IV(LVG), serial unknown, of *KG3*

Gotha G.IV, serial unknown, of *KG3*

Above: A Gotha G.IV, confirmed by the aileron actuating strut, takes off. It is painted in light colors for daylight bombing.

Below: This photograph of an operational G.IV being positioned the old-fashioned wasy was made into a Sanke card. Although the fuselage is still a light color, the wings and tail have been camouflaged in dark lozenge colors for night bombing raids over Britain. It is an interesting safety note that the aircraft is carrying bombs while being moved.

Above: An early Gotha G.IV painted in light colors for daylight bombing.

Above: An early Gotha G.IV, possibly 610/16, wears an over-size Gotha company trademark below the iron cross insignia on the rudder.

Above Right: The crater was made by a British bombing attack on the airfield, but *Lori 2*, flown by *Lt*. Kurt Küppers and *Oblt*. Fritz Lorenz of *Staffel* 16 seems undamaged other than for its missing engine nacelle panels, which apparently were blown off by the bomb's blast. With typical gallows humor, the sign reads "*Lori2* before the grave." The aircraft name is repeated on the rear fuselage. The aircraft is in dark camouflage for night bombing missions.

Gotha G.IV Test Aircraft

G.210/17: Had Maybach Mb.IVa engines, tractor propellers, and a 24.94 m wing span. Later re-engined with NAG C.III engines and pusher propellers.

G.211/17: Had Mercedes D.IVa engines and four-bladed tractor propellers.

G.212/17: Had a span of 28.20 m.

G.213/17: Had Flettner servo-controls on all control surfaces.

G.220/17: Had Flettner servo-controls on all control surfaces.

G.227/17: Was powered by 260 hp Mercedes D.IVa engines supercharged with an SSW compressor.

Above: A Gotha G.IV(LVG) in Austro-Hungarian service. The 39 Austro-Hungarian G.IV(LVG) bombers delivered were assigned Austrian serials 08.01–08.40; one had crashed in a "typical landing accident" by an LVG pilot and was not replaced. These aircraft had twin fuel tanks above the wing center section and were camouflaged in dark German hexagonal fabric for night bombing.

Below: Austro-Hungarian Gotha G.IV(LVG) 08.12 has drawn a crowd. The bomb racks are visible under the wing center section. Because German engine production was inadequate for German needs due to Allied numerical superiority and the Royal Navy's distant blockade, the Austrians had to provide their own 230 hp Hiero engines for their Gothas. The revised engine installation was a disaster; weak engine bearers and unsuitable propellers caused excessive engine vibration that led to piping leaks and failures of the structure and instruments. Despite great efforts by the maintenance staff, by late September the Austrian Gothas were virtually grounded and the aircrews reverted to the reliable, single-engine Brandenburg C.I for most bombing missions.

Gotha G.IV(LVG) 08.12 of *Flik* 102/G
of the Austro-Hungarian Air Service

Gotha G.IV *LoRi2* flown by *Lt.* Kurt Küppers
and *Oblt.* Fritz Lorenz of *Staffel* 16, *KG*3

Gotha G.IV, serial unknown, of *KG3*, Summer 1917

Gotha G.IV(LVG), serial unknown, of *KG3*, Summer 1917

Above: Austro-Hungarian Gotha G.IV(LVG) 08.16 of *Flik* 102/G was brought down by Italian anti-aircraft fire on 28 July 1918. The slot for the gunner to fire his gun downward through the "Gotha Tunnel" is clearly visible, as is the faired bulge behind it typical for the Gotha G.IV. The dark night camouflage, characterized in the resulting Allied report as "a very black, bluish gray", is evident.

Below: Gotha G.IV(SSW) G.211/17 was one of the aircraft used to evaluate design modifications. Here it is fitted with 260 hp Mercedes D.IVa engines mounted in tractor configuration and fitted with four-bladed propellers. In this installation the radiators were suspended from the upper wing.

Above: This late-production Gotha G.IV was covered in fabric printed with dark, hexagonal night camouflage for night bombing missions. Another detail of interest are the screens on both sides of the rear gunner's cockpit to prevent him from accidentally putting a hand or arm in the propeller arc.

Left: Gotha G.IV(LVG) 106/17 in a pose too often typical of a Gotha bomber.

Gotha G.V

The Gotha G.V was a further refinement of the basic design. Experience had shown that housing the fuel tanks in the engine nacelles increased the chance of fire in the event of a crash. Therefore the fuel tanks of the G.V were moved to the fuselage away from the engines, which were enclosed in streamlined nacelles mounted between the wings. The engine remained the same 260 hp Mercedes D.IVa as used in the G.III and G.IV. Despite the nominal improvements incorporated into the G.V, the basic design had reached its peak. Furthermore, wartime shortages resulted in inferior materials being used to construct the G.V, making it heavier than the G.IV and reducing its performance. Additional equipment deemed necessary by operational experience aggravated the problem.

Regardless, *Idflieg* ordered 100 G.V bombers on 19 October 1916. The G.V passed its type test in July 1917 and the first aircraft reached *Bogohl* 3 in August 1917. The G.V arrived too late to participate in the daylight raids over the UK but were used for night bombing, performing both strategic raids over the UK and tactical bombing on the Western Front.

The Gotha G.Va was an attempt to improve flight safety. A re-designed 'box' tail with biplane horizontal stabilizers and elevators and twin fins and rudders improved engine-out controllability, even enabling turns into the running engine. A two-wheeled *Stossfahrgestell* mounted under the nose prevented nose-overs on landing to improve landing safety, the Gotha's Achilles Heel. The nose turret was also modified. Twenty-five of this transitional model were delivered.

The G.Vb was derived from the G.Va in a further attempt to improve flight safety. The 'box' tail was retained and, to reduce the pilot's workload and

Gotha G.V Specifications

Engines:	2 x 260 hp Mercedes D.IVa	
Wing:	Span Upper	23.70 m
	Span Lower	21.90 m
	Area	89.5 m²
	Chord Upper	2.30 m
	Chord Lower	2.30 m
	Gap	2.22 m
	Sweepback	1.16°
General:	Length	12.40 m
	Height	4.30 m
	Empty Weight	2413 kg
	Loaded Weight	3648 kg
Maximum Speed:		140 km/h
Climb:	1000m	2.5 min
	2000m	8.5 min
	3000m	17 min
	4000m	29 min
Range:		840 km

Above: The distinguishing feature of the Gotha G.V was its new engine nacelles suspended between the wings. The fuel tanks were moved from the nacelles, their location in the G.IV, to the fuselage to reduce the danger of fire in a crash.

Above: This view of a Gotha G.V flown by *Kasta* 17 of *Bogohl* 3 shows its new engine nacelles, the night camouflage, and the original iron cross insignia over-painted to the new, straight-sided insignia. The bomber has an interesting marking of a constellation of eight-pointed stars with a line connecting them. Plates behind the wheels deflect debris from going through the propeller arc. A large load of bombs is carried beneath the wing center section.

improve maneuverability, most G.Vb bombers were equipped with Flettner servo controls on the upper ailerons. To further improve landing safety a four-wheeled undercarriage underneath each engine, similar to the original G.II prototype, was fitted. Unfortunately, the new undercarriage reduced the already compromised flight performance due to its greater weight and drag, and the Gotha G.Vb was inferior to the contemporary AEG G.V and Friedrichshafen G.IV based on comparative tests by *Idflieg*. Confirming this assessment, *Idflieg* placed an order with Gotha to build 50 Friedrichshafen G.IVa(Go) bombers under license, although the war ended before any were built. Although 80 Gotha G.Vb bombers were completed, the final few were delivered postwar directly to the Allies as part of the Armistice conditions.

Above: Front quarter view of an early Gotha G.V with faired struts supporting the engine nacelle.

Above & Below: Other than the change in location of the fuel tanks that affected the engine nacelles, the Gotha G.V was otherwise like the G.IV in all important respects. Not only did it offer no performance improvement, the deteriorating quality of the materials available for construction meant the G.V was a heavier airplane, reducing performance.

Above: Rear view of the Gotha G.V; this aircraft has only one over-wing gravity fuel tank.

Gotha G.IV(LVG) 991/16 *MoRoTas* of *Lt.* Mons, *Lt.* Roland, and third, unknown crewman of *KG3*, February–March 1918

Gotha G.V 925/16 *Pommern* of *Lt.* Rist, *Vzfw.* Gummelt, and *Vzfw.* Huhnsdorf of *Staffel* 15, *KG3*, 19–20 May 1918

Above, Below, & Bottom: Views of a night-camouflaged Gotha G.V in French hands with tri-color on the rudder undergoing evaluation. Until August 1918 when the Farman F.50 commenced operations the French had no twin-engine night bombers comparable to the Gotha in service; French night bombers were Voisin 8 and 10 single-engine pushers.

Above & Below: Gotha G.V 904/16 at the factory before night camouflage and the *Stossfahrgestell* were applied. The fairings on the engine nacelle struts were later removed. Seven 50-kg PuW bombs are under the center section.

Below: This Gotha G.V fitted with a simplified undercarriage is either the prototype or an early production model.

Gotha G.V, serial and unit unknown

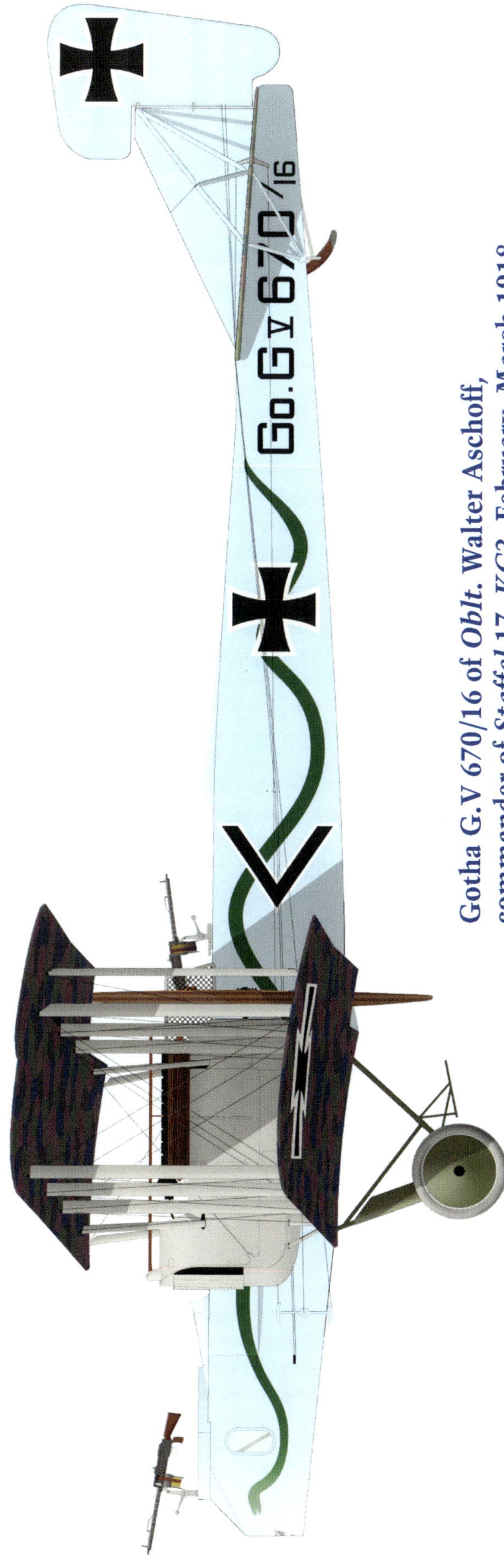

Gotha G.V 670/16 of *Oblt.* Walter Aschoff, commander of *Staffel* 17, *KG3*, February–March 1918

Above: Some Gotha G.V bombers were fitted with an eight-wheel *Stossfahrgestell* undercarriage to prevent nose-overs.

Above: To improve controllability with one engine out, the Gotha G.Va with 'box' tail was developed. Above is the G.Va prototype fitted with an early version of the box tail.

Below: The Gotha G.Va prototype with the final configuration of the box tail used in production.

Gotha G.V 979/16 of *KG3* was lost over England on the night of 19/20 May 1918. Of the crew of *Lt.* Flathow, *Vzfw.* Sachtler, and *Uffz.* Tasche, only Tasche survived the crash near Frinstead.

Gotha G.V 947/16 of *Lt.* von Korff, *Lt.* von Zedlitz, & *Gefr.* Speyer of *Bogohl* 3, March 1918

Above: Gotha G.Va and G.Vb bombers stored in a hangar. The two G.Vb bombers in the foreground are not fitted with the Flettner servo-assist on the ailerons, normally a feature of the G.Vb.

Above: The tail of Gotha G.Va 723/17 brought down in France on 5 July 1918 wears a dramatic winged dragon personal marking. This aircraft is the one in the upper right of the photograph at the top of this page.

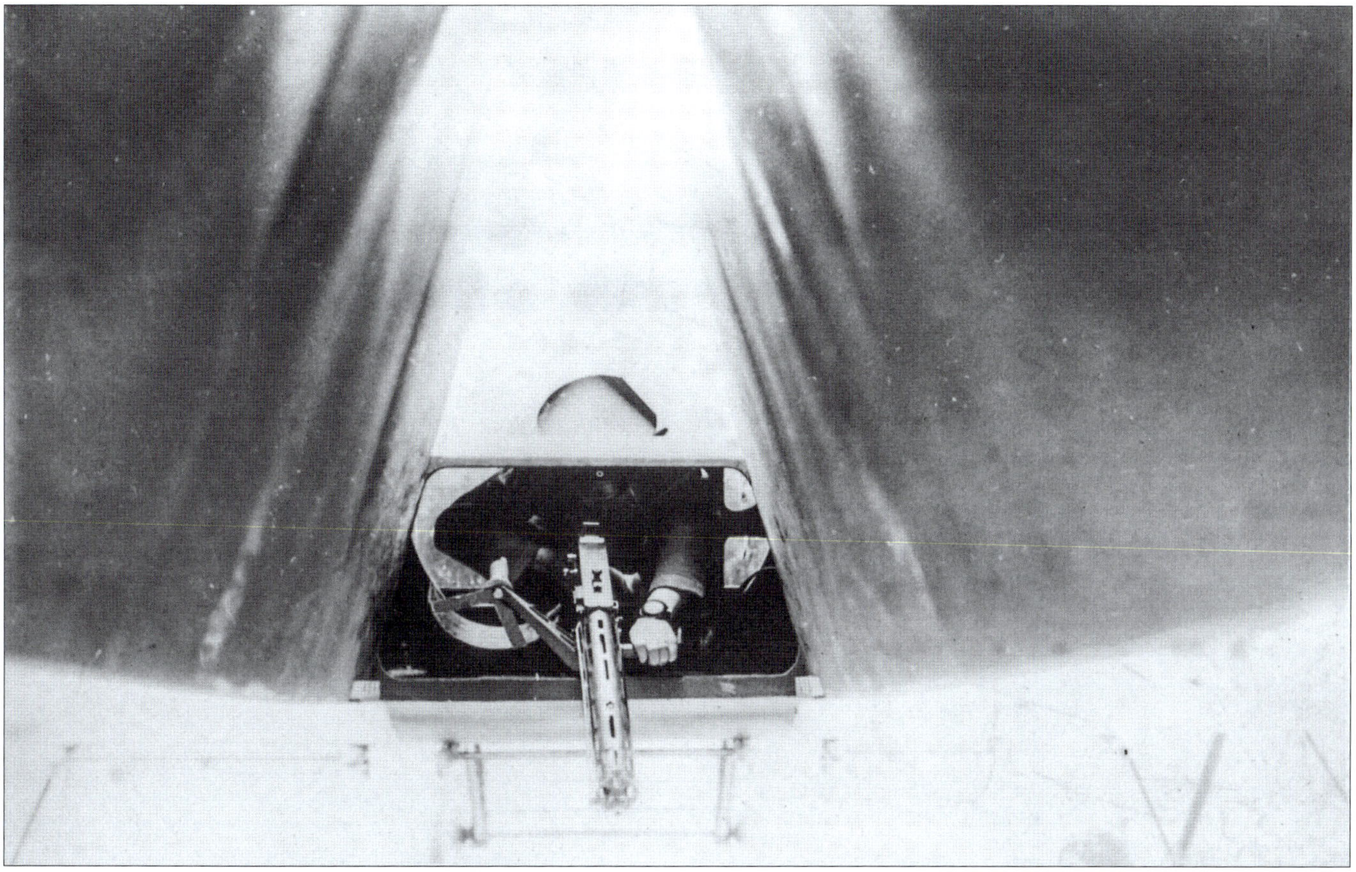

Above: The 'Gotha Tunnel' underneath the fuselage of a Gotha G.Va (notice the front undercarriage struts) as seen from the rear and being demonstrated by a gunner. The tunnel was introduced on the G.IV and carried forward into the G.V series; it enabled the rear gunner to protect the bomber against attacks from behind and below. A third machine gun could be mounted as shown, or the rear gunner's standard gun could fire downard through the slot in the rear fuselage as shown on the facing page.

Left: Gotha G.V in night camouflage.

Above & Above Right: Two more views of the tail of Gotha G.Va 723/17 brought down in France on 5 July 1918.

Above & Below: The nose undercarriage indicates this is likely a Gotha G.Va. These well-known photos show a gunner demonstrating the field of fire of the front and rear gunner's positions in a night-camouflaged bomber. Below the gunner demonstrates how to protect against an attack from below by firing down through the tunnel in the fuselage. The screens on the sides of the rear gunner's cockpit keep his hands out of the propeller arcs. A rare Gotha G.VI prototype is in the background in the photo above.

Gotha G.Va 723/17 of *Bogohl* 3, July 1918

Gotha G.Vb 917/18 *Hessen-Nassau*

Above & Below: Desperate to solve the landing accident dilemma that caused 76% of all Gotha bomber losses, the engineering team again modified the landing gear to create the Gotha G.Vb. The Gotha G.Vb reverted to an eight-wheel landing gear reminiscent of that fitted to the original G.II prototype. The additional forward-mounted wheels essentially eliminated the problem of nose-overs on landing, but at the expense of more weight and drag that reduced speed, range, and ceiling. This was a compromise the Gotha G.V could ill afford; it was already heavier than the G.IV due to inferior materials and more required equipment, and was now in an irreversible downward performance spiral.

Above: Gotha G.V *LoRi 3* of *Kasta* 16, *Bogohl* 3, and crew.

Below: Ground crewmen illustrate the range of bombs that Gotha bombers could carry with Gotha G.V 901/16 in the background. From left to right, the bombs are 12.5 kg, 50 kg, 100 kg, and 300 kg PuW bombs. Reproduced as a Sanke card, this was a popular propoganda photograph.

Above & Below: Ground crewmen load bombs by hand onto Gotha G.V 901/16.

Above: Gotha G.V 901/16 fully loaded with bombs; five 50-kg and two 100-kg PuW bombs are on the racks.

Below & Below Right: Groundcrew handling bombs to load on a Gotha. They are doing most of the work manually, with a primitive cart their only mechanical aid. These scenes could have been taken at any German base supporting twin-engine bombers in 1917 or 1918, regardless of the bomber's type.

Above: An impressive view of Gotha G.V 901/16 loaded and ready to start engines for its mission. The dark night camouflage shows the aircraft was normally used for night bombing, so the actual mission will not be flown until after dark. The Gothas ran most of their control wires outside the airframe, creating additional drag.

Above: Closeup of bombs hung under the extreme nose of a Gotha to provide a proper center of gravity. The Gothas were stable until the bombs were dropped, but pitch stability deteriorated significantly after that!

Above: The cockpit and flight controls of Gotha G.V 904/16 are typical for Gotha G.IV bombers as well. The cellon window above the instruments provided illumination during the day.

Above: Another view of a fully-loaded Gotha G.V. A number of 12.5 kg bombs were usually carried in the forward cockpit.

Below: With a Gotha G.V in the background, ground crewmen fill oxygen bottles with liquid oxygen for use by the bomber crew in flight. Photos on the following pages show how the crew sipped the oxygen in flight.

Leergewicht: 2740 kg.
Nutzlast: 1235 kg.
zul. Gesamtgewicht: 3975 kg.
Auf rechter Seite aufsteigen
2.563.

A Gotha crewman inflates his life jacket before a mission over the English Channel.

The Gotha crewman gets a final adjustment of his flight gear from a ground crewman before take-off.

Gotha G.VI

Chief Engineer Burkhardt had been thinking about how to reduce drag to improve speed, and thought that eliminating one of the engine nacelles would accomplish that despite the resulting asymmetric configuration. Burkhardt decided the most efficient way to design the aircraft was a tractor engine in the fuselage and a pusher engine in a nacelle to starboard which also had an observer/gunner's cockpit in the front that extended forward of the tractor propeller for greatest field of fire.

To compensate for asymmetric thrust and drag, the starboard nacelle was placed closer to the centerline of the wing than the fuselage. The thrust lines of the engines were close together, reducing asymmetric thrust in case of an engine failure. Burkhard obtained German Patent number 300 676 for his innovative design on September 22, 1915. *Idflieg* approved further study on 7 September 1915 but held off construction until design studies were completed. On 26 June 1916 Gotha management approved construction of the G.VI, which was underway the next month, but *Idflieg* did not sign a contract until 5 July 1917, when three G.VI bombers, serial numbers G.370–372/17, were ordered.

Gotha G.VI Specifications

Engines:	2 x 260 hp Mercedes D.IVa	
Wing:	Span Upper	23.70 m
	Area	89.5 m^2
General:	Length	12.36 m

The first Gotha G.VI was flight-tested in the fall of 1917 and was badly damaged in a landing crash in November. The two-bay wing was similar to that used in the Gotha G.II and the fuselage was derived from the G.IV. The second Gotha G.VI had three-bay wings and large radiators. Flight tests of this aircraft began in January 1918 and continued through March. However, flight performance was mediocre and there were many radiator problems as well. In April work on the G.VI was abandoned to allow Gotha engineers to focus on the Gotha GL.VII and GL.VIII prototypes.

The Gotha G.VI is thought to be the first asymmetric aircraft design to fly, proving the asymmetric conception was practical. Although Burkhardt claimed that G.VI performance exceeded its conventional predecessors and that it flew very well, his description of its flying qualities may be optimistic given its crash-landing.

Right & Facing Page: The first Gotha G.VI prototype had two-bay wings. It is shown here after a landing accident in November 1917. The fuselage and tail were derived from the Gotha G.IV and the wings were similar to those of the Gotha G.II.

Below: The second Gotha G.VI prototype had larger, three-bay wings. Flight tests began in January 1918 and continued through March. In April work on the G.VI was abandoned due to mediocre flight performance. Problems with the radiators were a contributing factor. Furthermore, terminating G.VI development enabled Gotha engineers to focus on development of the more promising GL.VII and GL.VIII bombers.

Gotha GL Types

Realizing that the standard Gotha G series was too vulnerable to interception, *Idflieg* released a specification for a high-performance G-type, designated GL type (GL for lightened G-type). The specification required daylight operation at 6–7,000 meters while carrying 300 kg of either bombs or photographic equipment.

The initial aircraft were designed by Rösner and Schleiffer and were completely different from the earlier Gotha bombers.

The Gotha GL ('GL' meant *erleichtertes-G flugzeug*, or lightened G aircraft) series were intended to meet the *Idflieg* requirement for long-range, high-speed aircraft for reconnaissance and bombing. In official records the designations were GL.VII, GL.VIII, and GL.IX, but the simple G-designation was used interchangeably. This has lead to some confusion, but the G.VII and GL.VII were the same aircraft, as were the G.VIII and GL.VIII and the G.IX and GL.IX.

Gotha G.VII/GL.VII

Ordered on 15 March 1917, before daylight operations by Gotha G.IV aircraft had begun over the UK, four prototypes were built. These were designated Gs.1 by Gotha and G.VII and G.VIII by *Idflieg*. The first prototype, Gotha GL.VII 550/17, was completed in September 1917, the month that *Kagohl* 3 transitioned to night bombing of the UK due to strengthened British defenses and increasing losses of Gotha G.IV bombers on daylight missions.

The single-bay Gotha GL.VII 550/17, powered by two 260 hp Mercedes D.IVa engines, was one of four prototypes (G.550–553/17) ordered at the same time to explore different engine and wing configurations to determine the optimum layout to meet the demanding *Idflieg* requirements. A characteristic of the design was the close spacing of the engines to minimize asymmetric thrust with one engine out. This configuration was based on a patent by *Ingenieur* Michael Schlieffer, a Gotha pilot, and used the venturi effect between the nacelle and fuselage to minimize drag and help counter asymmetric

Gotha G.VII/GL.VII Specifications

Engines:	2 x 260 hp Mercedes D.IVa	
Wing:	Span	19.27 m
General:	Length	9.60 m
	Empty Weight	2420 kg
	Loaded Weight	3140 kg
Maximum Speed:		180 km/h
Service Ceiling:		6000 m

Above: The prototype of a Gotha GL-type was the single-bay G.VII G.550/17, shown in September 1917. The engines rotated in opposite directions to minimize torque and the wheels are faired to reduce drag. However, the fuselage is suspended between the wings by a complex set of drag-producing struts.

Above: Gotha G.VII G.550/17 as built was very compact for a twin-engine aircraft.

Below: A Gotha GL-prototype, probably G.VII G.550/17 due to its wheel fairings, displays its very clean lines. Early GL-types had a single fin and rudder.

Above: Gotha G.VII G.550/17 shown in September–October 1917 with wheel fairings removed for flight tests.

thrust. The close engine spacing required a short nose that precluded a forward gunner.

The next prototype, Gotha G.VII 551/17, had a longer, two-bay wing cellule. The Gotha G.VII was faster than the Albatros D.III but, despite months of intensive testing to optimize engine, airframe, and propeller configurations to maximize speed, climb, and ceiling, the desired performance was not achieved. In late 1918 engines with experimental centrifugal superchargers were installed in this machine for testing.

The production standard GL.VII had a two-bay wing cellule with aerodynamically balanced ailerons on the upper wing and a box (biplane) tail. In 1918 eight aircraft were produced from an order for 55 aircraft (G.300–354/18). Gotha GL.VII 300/18 was the first of these and was powered by two Maybach Mb.IVa engines. It was sent to the front for evaluation in May 1918. GL.VII 302/18 went to the front on 27 June, followed by 301/18 on 1 August and 304/18 on 26 October. However, no further information is available on the evaluation of these aircraft and it is not known if they flew any operational sorties.

Owing to promising results obtained in early Gotha tests, both Aviatik and LVG were awarded contracts in February 1918 to build the Gotha G.VII under license. Aviatik built 30 G.VII aircraft (G.100–129/18), few of which were handed over to the Allies after the war. Aviatik-built Gotha G.VII(Av) 112/18 was experimentally fitted with nose radiators for flight trials.

Below: One of the Gotha GL-prototypes, still with a single fin and rudder, now has the fuselage faired into the lower wings to reduce drag.

Above: This Gotha GL-prototype may be the same aircraft as the one at the bottom of the previous page, but the propeller spinners have been removed.

Below: A Gotha GL.VII G.300/18 represents the production version of the aircraft. Now it has a 'box' tail for better directional control with one engine out, no wheel fairings, and no propeller spinners. The wings are now longer, with two-bays of struts, and the aileron aerodynamic balances have slots in them.

Above: A rear quarter view of Gotha GL.VII G.300/18, the production version of the GL.VII. The 'box' tail is visible and the rudders, like the ailerons, have aerodynamic balances with slots in them.

Below: This Gotha GL.VII in the factory has no markings, but the wings are covered in regular camouflage fabric with hexagons. The fuselage has been spray-painted in hexagons with soft edges. This may be one of the GL-prototypes. The 'box' tail has the more common aerodynamic balances without slots in them.

Above: Aviatik-built Gotha GL.VII(Av) G.112/18 was fitted with nose radiators for evaluation. The aileron aerodynamic balances do not appear to have slots. Aviatik built 30 GL.VII bombers.

Below: This Gotha GL.VII has an aluminum nose and propeller spinners. The radiators were mounted in the upper wing above the engines, the position used for production aircraft.

Above & Below: Post-war Ukrainian Gotha GL.VII *Olena* made an emergency landing at Vajnory airfield near Braislava, Czechoslovakia, in early 1920. Despite the fact that Ukrainian aircraft had been over-flying Czechoslovakia with permission since 1919, Czechoslovakian authorities interned *Olena*. After repairs *Olena* was flown to Prague for evaluation in February or March. The German pilot flew the aircraft, but he had two Czech mechanics with him.

Above: This Gotha GL.VII *may* have been photographed in Czechoslovakia, but details are unknown.

Above: Postwar lineup of a Gotha GL.VII (at left), an AEG (center), and a Friedrichshafen (at right) at Bickendorf. The GL.VII has a fuselage band over the cross insignia, its location indicating it was built by Aviatik.

Right: Gotha GL forward fuselage at the USAS center at Romorantin, France postwar.

Gotha G.VIII/GL.VIII

Intended for high-speed day bombing, the Gotha GL.VII was fitted with an enlarged, three-bay wing to improve climb and ceiling. One fixed, forward-firing machine gun was fitted for the pilot, and the gunner had both dorsal and ventral flexible guns for defense.

The Gotha GL.VIII prototype, G.VIII 552/17, was powered by Mercedes D.IVa engines, but the production standard was the over-compressed 260 hp Maybach Mb.IVa for more power at high altitude. All control surfaces were aerodynamically balanced and the biplane tail was retained for improved control with one engine out. The PuW bombs were carried externally under the fuselage.

Gotha GL.VIII 307/18, powered by two 260 hp Maybach Mb.IVa engines, was sent to the Döberitz test center on 19 October 1918. The upper ailerons were fitted with Flettner servo controls for lighter control forces. According to an *Idflieg* appreciation, the performance of these aircraft "lagged far behind expectations." On the other hand, the war ended before sufficient experience was gathered to evaluate the aircraft adequately under combat conditions.

Gotha G.VIII/GL.VIII Specifications

Engines:	2 x 260 hp Maybach Mb.IVa	
Wing:	Span	21.73 m
	Chord	2.05 m
	Gap	2.00 m
General:	Length	9.79 m
	Useful Load	1030 kg
	Loaded Weight	3706 kg
Maximum Speed:		180 km/h
Service Ceiling:		6000 m

Below: Although some Gotha GL prototypes appeared to carry a fixed gun for the pilot, the Gotha GL.VIII high-speed day bomber was the only type for which a fixed machine gun for the pilot was specified. Here the gunner demonstrates the field of fire of his dorsal gun; the fixed pilot's gun is also clearly visible. The gunner also had a flexible ventral gun to defend against attacks from below. The lozenge printed fabric and clean lines are evident.

Above & Below: Gotha GL.VIII G.307/18, a production GL.VIII, was powered by two 260 hp over-compressed Maybach Mb.IVa engines optimized for power output at high-altitudes for maximum ceiling and speed at altitude. The slanting struts supporting the upper wingtips characterize the production GL.VIII and differentiate it from the GL.VII.

Gotha GL.VIII 307/18

Gotha G.IX/GL.IX

In February 1918 LVG was given a production contract to build 30 Gotha GL.VII reconnaissance airplanes (G.200–229/18) and 70 Gotha G.IX bombers (G.230–299/18). The 260 hp Maybach Mb.IVa was the normal powerplant. Apparently all of the GL.VII(LVG) aircraft were built but no photographs of this version have been found. The G.IX day bomber, given an increased wing-span and length, was equipped to carry five 50 kg PuW bombs under the fuselage. According to a French inspection report, some 96 Gotha G.IX bombers were stored at the LVG factory on 15 December 1918. After the armistice a number of these bombers were given to the Allies. For example, Belgium received 23 G.IX bombers that were flown operationally by the Belgian air service. A Gotha G.IX(LVG) 289/18, delivered to Japan post-war as reparations, was powered by two 260 hp Mercedes D.IVa engines.

Gotha G.IX/GL.IX Specifications

Engines:	2 x 240 hp Maybach Mb.IVa

Above: Gotha G.IX(LVG) G.263/18 displays the LVG-style markings and is armed with a PuW bomb under the fuselage.

Below: Gotha G.IX(LVG) G.257/18 is covered in dark, printed camouflaged fabric.

Above: This Gotha G.IX(LVG) illustrates the extended wing-span of this version of the GL series. A British crew attempted to fly this aircraft back to Britain, but it crashed when leaving Bickendorf and the crew was killed.

Below: This Gotha G.IX(LVG) illustrates the type's clean lines. It is serving in Belgium postwar. The number '250' on the nose *may* indicate it is serial G.250/18, but there is no confirmation. Roundels are being painted under the wings but are not yet complete. The aircraft is covered in dark, printed camouflage fabric.

Above & Below: Gotha G.IX(LVG) in Belgian service postwar. These views illustrate details and emphasize the aerodynamic lines of the GL series. Problems of propeller selection for maximum ceiling and speed at altitude, always a challenge with fixed-pitch propellers, prolonged development of the GL-series.

Above: Gotha G.IX(LVG) in Belgian service postwar.

Below: The Gotha G.IX(LVG) in this postwar view has no visible markings other than the Belgian colors on the rudders, but its hexagonal camouflage fabric is well illustrated. The aileron aerodynamic balances have slots, but the aerodynamic balances on the rudders do not. A DH.9 is visible in the background.

Above: This Gotha G.IX(LVG) 299/18 illustrates its LVG-style markings, its camouflage fabric, and the fairings over the radiators in the upper wing above the engines.

Below: Gotha G.IX(LVG serving in Belgium postwar illustrates the type's clean lines. It appears to be on public display. A Friedrichshafen bomber is in the background.

Above & Below: Gotha G.IX(LVG) in Belgian service postwar.

Gotha GL.VII 300/18

Gotha G.IX(LVG) in Postwar Belgian Service

Gotha G.X/GL.X

On 16 August 1918, *Idflieg* ordered three Gotha GL.X prototypes (G.1725–1727/18), each powered by two 185 hp BMW.IIIa engines. Designed by Burkhardt, two GL.X versions were envisioned, one a fast, high-altitude reconnaissance machine and the second an armored ground-attack aircraft fitted with a single, forward-firing 20mm Becker cannon and five fixed machine guns. A gunner had a sixth, flexible machine gun for rear defense. The first two-bay GL.X, a more compact design than the G.IX bomber, was reported ready for flight testing on 1 January 1919, subject to clearance from Gotha management. It is difficult to understand how the same engine and airframe combination was expected to perform well at both extreme high and low altitudes.

Gotha G.X/GL.X Specifications	
Engines:	2 x 185 hp BMW.IIIa

Above: The Gotha G.X prototype had two-bay wings and was powered by two 185 hp BMW.IIIa engines. Somehow this airframe and engine combination was supposed to excel as both a high-altitude reconnaissance aircraft and a low-level ground-attack aircraft, seemingly contradictory requirements. The massive WD27 looms in the background.

Above: Gotha GL fuselages at the USAS center at Romorantin postwar show the fuselage cross section.

Left: Gotha GL.IX in Belgium postwar.

Gotha Aircraft in Retrospect

Gotha's initial forays into aircraft production met with only modest success. The first Gotha aircraft were a small series of *Taube* types built in small numbers. Although the *Taube* configuration was obsolete before the war started, aircraft of any type were valuable for observation and reconnaissance, and the well-built, robust Gotha *Tauben* were among the best of this obsolete breed.

Following their *Tauben*, Gotha designed and built an undistinguished series of single-engine biplanes for land and sea. Of the seven Gotha land biplane types, five of which were original designs, only two merited small-quantity production as B-types and were mostly used for training. Gotha's single-engine seaplanes faired little better. Too fragile for the demands of operational use in the Baltic and North Sea, the few produced were either used for training or sent to Turkey, where the sea and weather conditions were not as demanding.

Originally a minor aircraft manufacturer of single-engine airplanes, Gotha started its transition into becoming a successful designer and builder of twin-engine and multi-engine warplanes when it signed a production license for the Gotha G.I based on Ursinus's design for the Friedal-Ursinus twin-engine battleplane. Gotha built 18 Gotha G.I aircraft in three production batches, but the battleplane concept, rooted in the experience of naval combat, was fundamentally flawed. Although not a great operational success, the Gotha G.I gradually morphed into a bomber, the natural role for a large, relatively slow twin-engine aircraft.

From the experience gained building the twin-engine G.I designed by Ursinus, Gotha next developed the Gotha G.II twin-engine bomber that was a completely new aircraft. Gotha produced only a few G.II bombers and hardly more G.III bombers. However, these transitional aircraft lead to the Gotha G.IV, a truly competitive bomber. The G.IV was the breakthrough Gotha needed. In fact, Gotha was little-known until the Gotha G.IV bombed London in broad daylight on 13 June 1917. This raid on the heart of the world-spanning British Empire shocked the British public and overnight made Gotha a household name around the world. From then on, the British public thought of all twin-engine German bombers as Gothas, regardless of their actual manufacturer.

The Gotha raids on Britain were the second stage of a strategic bombing campaign that Germany had been waging against Britain for more than two years with Zeppelins. At first Britain regarded the Zeppelin menace with dread, but over time the airships, bombing only at night for safety, had proven ineffectual. After a couple were downed over Britain in spectacular fashion, the British public started to dismiss the airships as a serious threat, many holding the airships in contempt.

The Gotha raids instantly changed that. For one thing, unlike the night-raiding Zeppelins, the Gothas bombed London in daylight and were witnessed by thousands of citizens. The ease with which the Gothas cruised over the British capital and bombed targets at leisure shocked the public. Furthermore, the first Gotha raid caused more casualties than any other bombing raid during the entire war. Finally, the total inability of British defenses to interfere with the Gothas, let alone bring any down, further enraged the public.

Recognizing the negligible military effect of the raids, the British government wanted to ignore them in favor of focusing on the crucial Western Front. But an outraged public would not tolerate that, and the most important direct impact of the Gotha raids was to compel the British government to greatly strengthen London's defenses with heavy guns and fighter squadrons desperately needed at the front. A longer-term effect was the establishment of the RAF, the world's first independent air force, by combining the RFC and RNAS, with the intent of improving the air defense of Britain.

Despite the huge emotional impact of the Gotha daylight raids on London and other British targets, greatly improved British defenses soon forced the Gothas to abandon daylight raids for night bombing.

A crucial design decision during development of the Gotha G.II, the direct ancestor of all Gotha bombers that reached operational service, had a pernicious effect on the Gotha bombers. That decision came during a revision of the G.II landing gear. The center of gravity was deliberately moved aft to allow a simpler landing gear to be fitted that would enable the tail skid to be used as a brake to stop the aircraft after landing. This simple modification, done for good reason, had a devastating impact on Gotha losses on operations. The change made the Gotha bombers too tail heavy, and thus marginally stable in pitch. During missions some bombs were mounted well forward to balance the aircraft on its outbound flight, but once the bombs were dropped the bombers' natural tail heaviness predominated, resulting in numerous landing accidents. In fact, landing accidents were responsible for 76% of all Gotha bomber losses, more than three times the amount for all other causes combined!

As landing accidents rapidly mounted, the problem and its solution should have been obvious. But airplane design is the art of practical compromise; improving one facet of an airplanes' operation often degrades another. Apparently still worried about stopping the bombers after landing without brakes, the center of gravity was not shifted forward again. Instead, the designers struggled with additional wheels forward of the main wheels, with the intention of preventing nose-overs on landing. The fundamental tail heaviness was not solved, and the greater weight and drag of the additional wheels reduced speed, climb, and ceiling.

This problem combined with another to drive Gotha performance into a downward spiral that resulted in obsolescence. Under the stress of war combined with the Royal Navy's distant blockade, the German economy was being strangled and material quality dropped. For Gotha the result was heavier airframes with no compensating power increases or aerodynamic improvements to maintain performance. While this was not as critical for a night bomber, the deteriorating performance of the Gotha bombers due to poor quality materials and unfortunate landing gear modifications decreased speed, ceiling, range, bomb-load, and overall flight safety. Eventually, Gotha was ordered to produce Friedrichshafen bombers under license.

In parallel with twin-engine bomber development, Gotha developed twin-engine torpedo planes. These seaplanes made a contribution to the aerial torpedo force, but experience soon showed that torpedo operations were not very effective. Worse, the ungainly torpedo bombers, required by the nature of their weapons to attack low and slow on a constant heading, were highly vulnerable to shipboard anti-aircraft fire. The result was that the German Navy soon abandoned aerial torpedo operations. Having a sizable inventory of large seaplanes designed for torpedo bombing, the Navy decided to re-purpose them as long-range maritime reconnaissance airplanes, and the Gotha torpedo planes were modified for that purpose. However, they were under-powered, partly due to the effects of the distant blockade on engine availability, and were also fragile, breaking up in any but the mildest sea conditions when afloat. These problems greatly limited their usefulness and they flew only limited numbers of reconnaissance sorties.

To regain the ability to perform long-range bombing in daylight, Gotha developed the GL series of bombers and long-range reconnaissance airplanes. Despite significant potential, Gotha was unable to develop these designs to achieve the desired speed and ceiling to enable them to avoid interception by Allied fighters. Although several aircraft were evaluated at the front, the GL-types did not see general operational service. Again, a major contributing factor was inadequate engine power due to a toxic combination of *Idflieg's* feckless engine development policies and strangulation of German industry by the Allied distant blockade.

Overall Gotha was not a greatly successful aircraft manufacturer, but had a brief moment of glory, or at least notoriety, when its G.IV bombed London in daylight. Forced to abandon daylight bombing missions in the face of greatly improved defenses, Gothas continued night bombing raids until the 19/20 May 1918. These raids tied down a large amount of critical resources that could have been better employed on the Western Front, but otherwise did no great damage of military significance.

Gotha was unable to follow its G.IV bomber with new designs that were successful on operations. After the Armistice Gotha exited the aviation business until a later war, having earned a reputation that exceeded its actual military accomplishments.

Gotha Bomber Attack Altitudes

Raid	Attack Altitude
1st Daylight	4,900–5,100 m (16,000–16,700 ft.)
2nd Daylight	4,500–4,900 m (14,700–16,100 ft.)
3rd Daylight	3,800–4,100 m (12,400–13,400 ft.)
1st Night	2,000–2,500 m (6,600–8,200 ft.)
Last Night	1,200–1,700 m (4,000–5,600 ft.)

Day: As the daylight raids progressed, Gotha bomber performance deteriorated as shown by the decreasing attack altitudes. There were several causes for this, including increased equipment carried (extra fuel tanks, oxygen, navigation equipment, messenger pigeons), increasing airframe weight as wood quality decreased, fuel quality deterioration, and engine and airframe wear.

Night: Gotha attack altitudes during the night raids were far lower than during daylight. One reason is that night raids did not need to be made at high altitude to avoid interception and anti-aircraft fire, and climbing to a lower altitude saved fuel and engine stress. In addition, use of *Stossfahrgestell* undercarriages on many night bombers, with their greater weight and drag, also affected aircraft performance.

Gotha G-Type Bomber Production Summary				
Order Date	**Type**	**Qty**	**Serials**	**Notes**
April 1, 1915	G.I	6	9–14/15	Plus one Friedel-Ursinus prototype, B.1092/14. Production G.Is delivered 27 July–8 Sep. 1915.
July 15, 1915	G.I	6	40–45/15	Delivered 22 Sep.–5 Nov. 1915.
Oct. 10, 1915	G.I	6	100–105/15	Delivered 24 Jan.–20 March 1916.
Dec. 18, 1915	G.II	10	200–209/16	Delivered Aug.–Sep. 1916
May 3, 1916	G.III	25	376–399/16	Delivered 16 Oct. 1916–26 March 1917
Aug. 4, 1916	G.IV	12	401–412/16	
Oct. 19, 1916	G.IV	25	600–624/16	
Nov. 23, 1916	G.IV	15	649–663/16	
Dec. 4, 1916	G.IV(LVG)	50	980–1029/16	
Dec. 18, 1916	G.IV(SSW)	40	1055–1094/16	Delivered July 1917–February 1918
May 1917	G.IV(SSW)	40	200–239/17	Delivered Dec. 1917–Aug. 1918
Aug. 1917	G.IV(LVG)	50	100–149/17	
Oct. 19, 1916	G.V	20	664–683/16	Delivered Aug. 1917–April 1918
Oct. 19, 1916	G.V	80	900–979/16	Delivered Aug. 1917–March 1918
Oct. 15, 1917	G.Va	25	700–724/17	Delivered 3 Apr.–22 May 1918
Oct. 15, 1917	G.Vb	25	725–749/17	Delivered 4 June–15 Aug. 1918
May 18, 1918	G.Vb	25	913–927/18	Delivered 30 June–6 Sep. 1918
July 26, 1918	G.Vb	15	1450–1464/18	Delivered 9 Sep.–2 Oct. 1918
Aug. 4, 1918	G.Vb	15	1778–1792/18	Delivered 18 Oct.–14 Dec. 1918
July 5, 1917	G.VI	3	370–372/17	Only first two built; third cancelled.
In addition to these aircraft delivered to *Idflieg* orders, LVG built an additional 40 G.IV bombers that were delivered to an Austro-Hungarian order and used 230 Hiero engines built in Austria. The squadrons received these aircraft, Austro-Hungarian serials 08.01–08.40, in March–April 1918.				

Gotha GL-Type Production Summary				
Manufacturer	**Type**	**Qty**	**Serials**	**Notes**
Gotha	G.VII & G.VIII	4	550/17 – 553/17	Prototypes; all 4 completed.
Gotha	G.VII & G.VIII	55	300/18 – 354/18	Production aircraft; only 8 were completed.
Aviatik	G.VII(Av)	100	100/18 – 129/18*	At least 30 completed.
LVG	G.VII(LVG)	30	200/18 – 229/18	Reconnaissance aircraft. All 30 believed to have been completed.
LVG	G.IX(LVG)	70	230/17 – 299/18	Bombers. Unknown number, perhaps all, completed.
Gotha	G.X	3	1725/18 – 1727/18	Prototypes. At least one completed.
Notes: 1. *Rest of serials unknown. 2. According to a French inspection report 96 Gotha G.IX aircraft were stored at the LVG factory on 15 Dec. 1918. Many of these were transferred to the Allies postwar, especially Belgium.				

Frontbestand Inventory of G-Type Twin-Engine Bombers at the Front

Manufacturer and Type		1914			1915						1916						1917						1918			
		31 Aug	31 Oct	31 Dec	28 Feb	30 Apr	30 Jun	31 Aug	31 Oct	31 Dec	28 Feb	30 Apr	30 Jun	31 Aug	31 Oct	31 Dec	28 Feb	30 Apr	30 Jun	31 Aug	31 Oct	31 Dec	28 Feb	30 Apr	30 Jun	31 Aug
AEG	G.I							1	5			5														
	G.II						2	5	10	13	12	2	4	4	4	2	2	1	1							
	G.III												6	16	22	21	22	9								
	G.IV									1								5	9	15	15	35	37	54	74	51
	G.																	3								
Albatros	G.II																1	9	1	1		1				
	G.III																		2	1	2					
Friedrichshafen	G.II														1		4	8	17	17	10	9	2	1	1	1
	G.III																		9	32	24	57	69	96	74	24
	G.IIIa																								18	95
	G.IV																								4	8
	G.IVa																								5	6
Gotha	G.I								5	6	1		1	1	1											
	G.II														4	3	1	1								
	G.III														7	14	3	4	3	3						
	G.IV															1		30	36	34	35	19	10	8	6	5
	G.V																			3	20	33	34	36	15	8
	G.Va																							11	19	4
	G.Vb																									21
Rumpler	G.I												1		1											
	G.II												1	7	8	4	1	1								
	G.III															1		3	5	5	10	1	4			
Total:							2	6	20	20	13	7	13	28	48	46	34	71	86	111	116	155	156	206	216	223

Twin-Engine Bombers (G Types)

Early in the war Germany pursued the '*Kampfflugzeug*' or battleplane concept, which lead to development of the twin-engine AEG K.I, the 'K' standing for '*Kampfflugzeug*'. The *Kampfflugzeug* was envisioned as a sort of aerial cruiser with gunners engaging enemy aircraft with flexible guns in a manner analogous to naval combat. Operational experience quickly revealed that the slow *Kampfflugzeug* was better suited to bombing than intercepting other aircraft and the designation of the AEG K.I was changed to AEG G.I, 'G' standing for '*Grossflugzeug*', literally 'large aircraft' but in this case also having the connotation of twin-engine bomber. AEG were first to field G-types, soon followed by Gotha, then Rumpler, Friedrichshafen, and Albatros.

Both the Rumpler and Albatros designs were mediocre and built in very small numbers. Contrary to the *Frontbestand*, the Albatros G.II remained a prototype. Only the Albatros G.III was produced and served at the front, and the inventory of G.II aircraft were actually Albatros G.III aircraft; the numbers in these two rows should be combined under G.III.

The Gotha went on to fame for bombing London in broad daylight in June 1917, and later switched to night bombing after British defenses were greatly strengthened, then bombed tactical targets behind Allied lines during and after the great German offensive in March 1918. In early 1918 Gotha abandoned development their existing designs in favor of smaller, faster designs, but despite prolonged engineering effort these did not reach the front.

The AEG G-types were successful tactical bombers from their appearance at the front. The AEG G.IV listed in December 1915 was a transcription error; the G.IV did not reach the front until 1917. Friedrichshafen G-types were also successful, both they and the AEG series being superior to the famous Gotha G-types in all respects except range and ceiling.

The G-types were very effective tactical night bombers and the armistice specifically required the Germans to surrender all night bombers to the Allies, who were astonished at the small number of aircraft they received given the damage they had done. Resource shortages, especially engines, limited the number of G-types Germany could build.

Gotha LE3

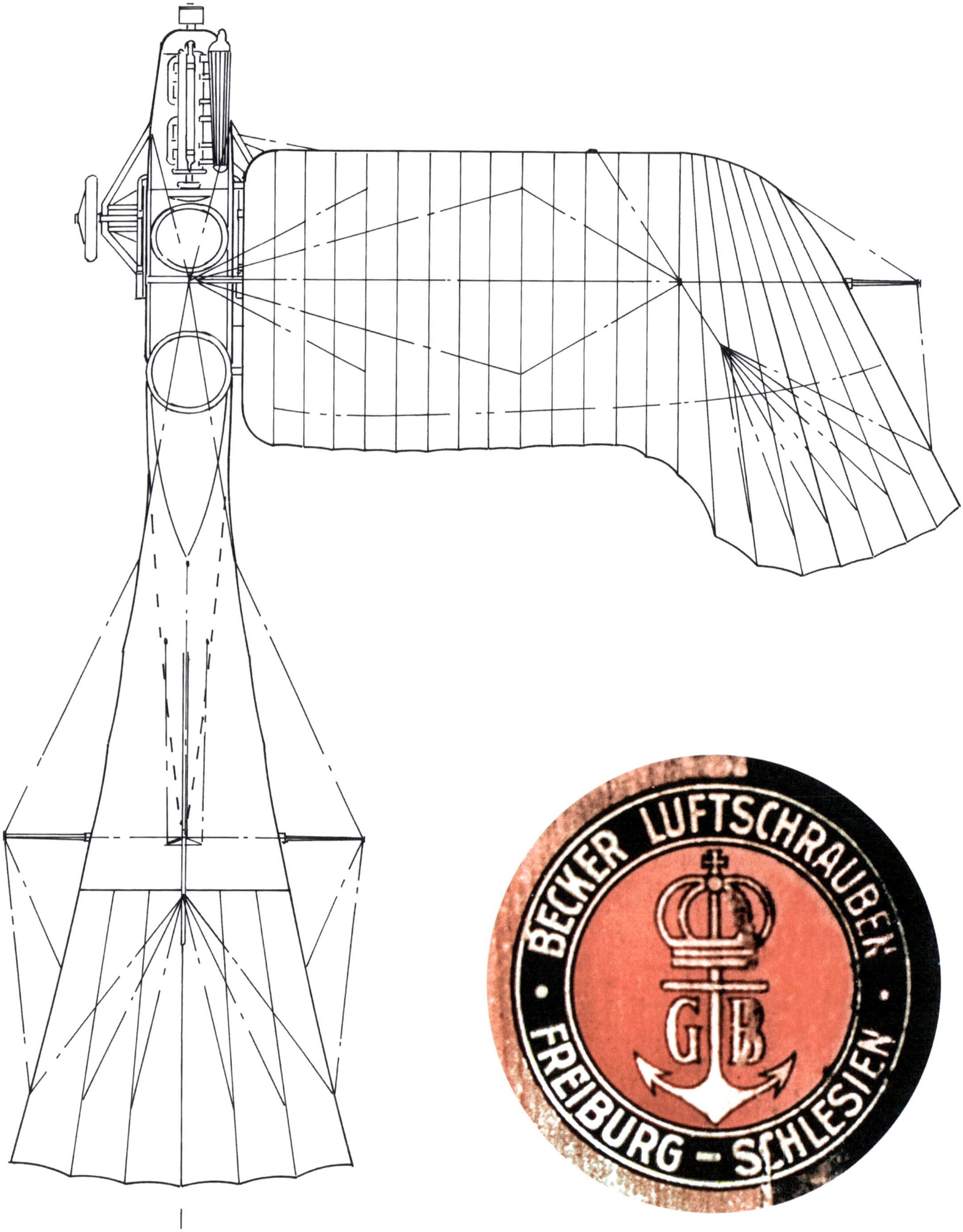

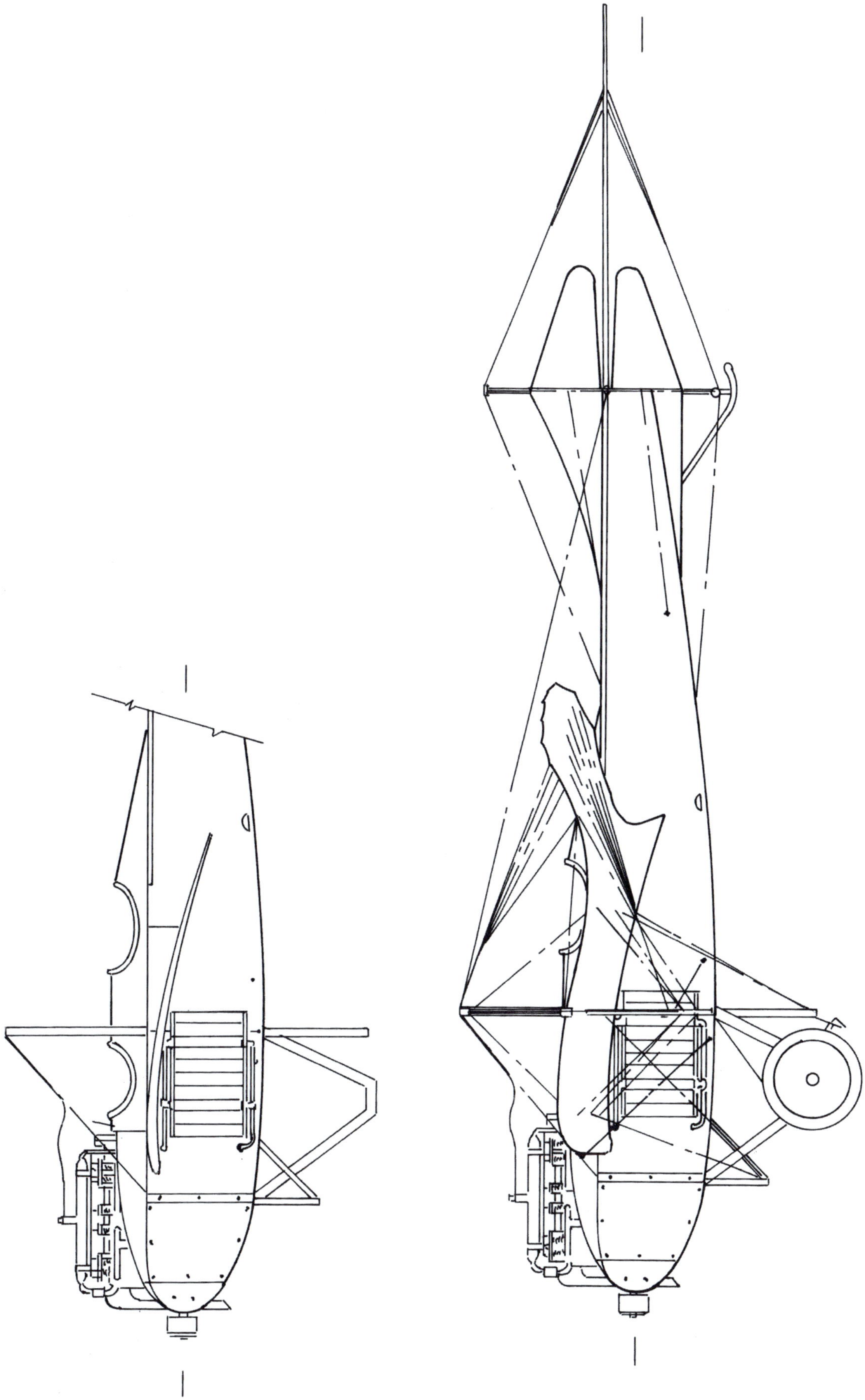

Gotha LE3 & LE4

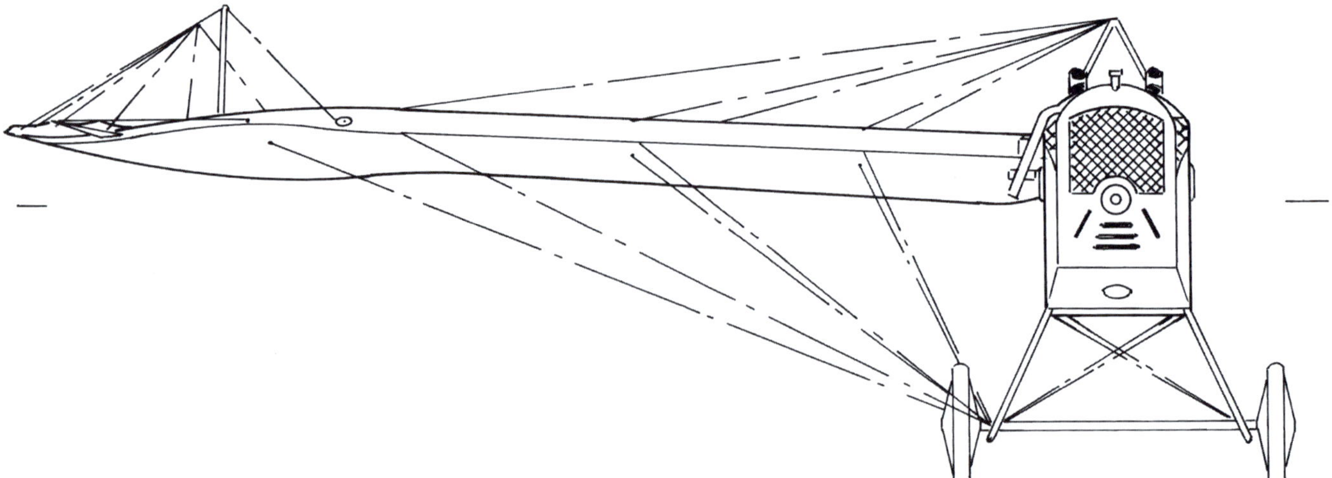

Gotha LE4

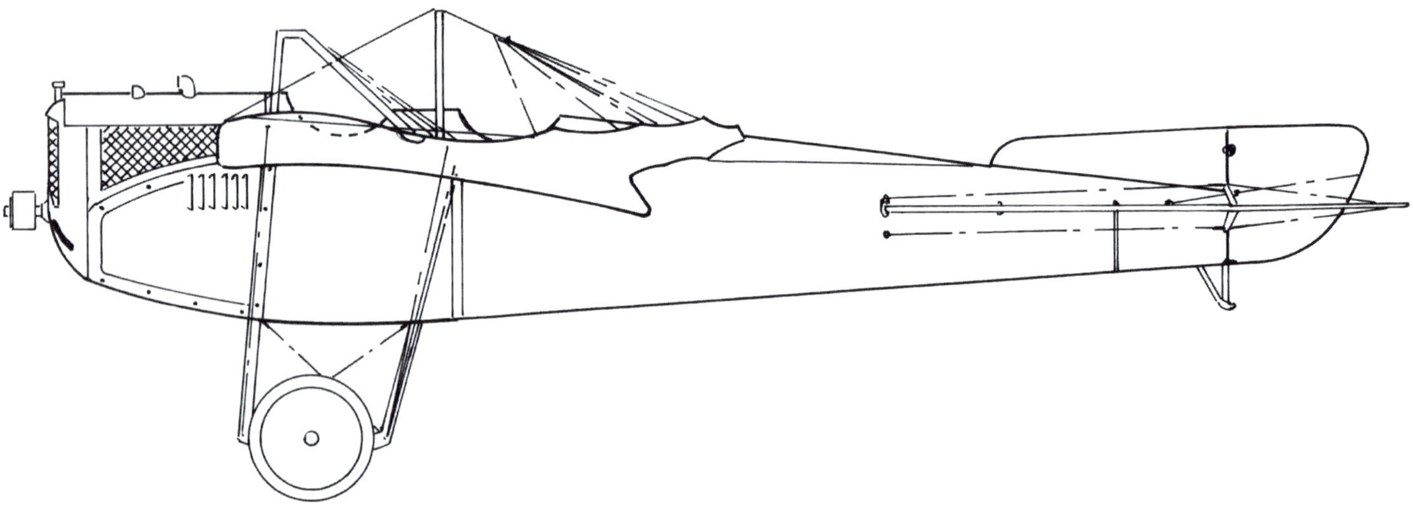

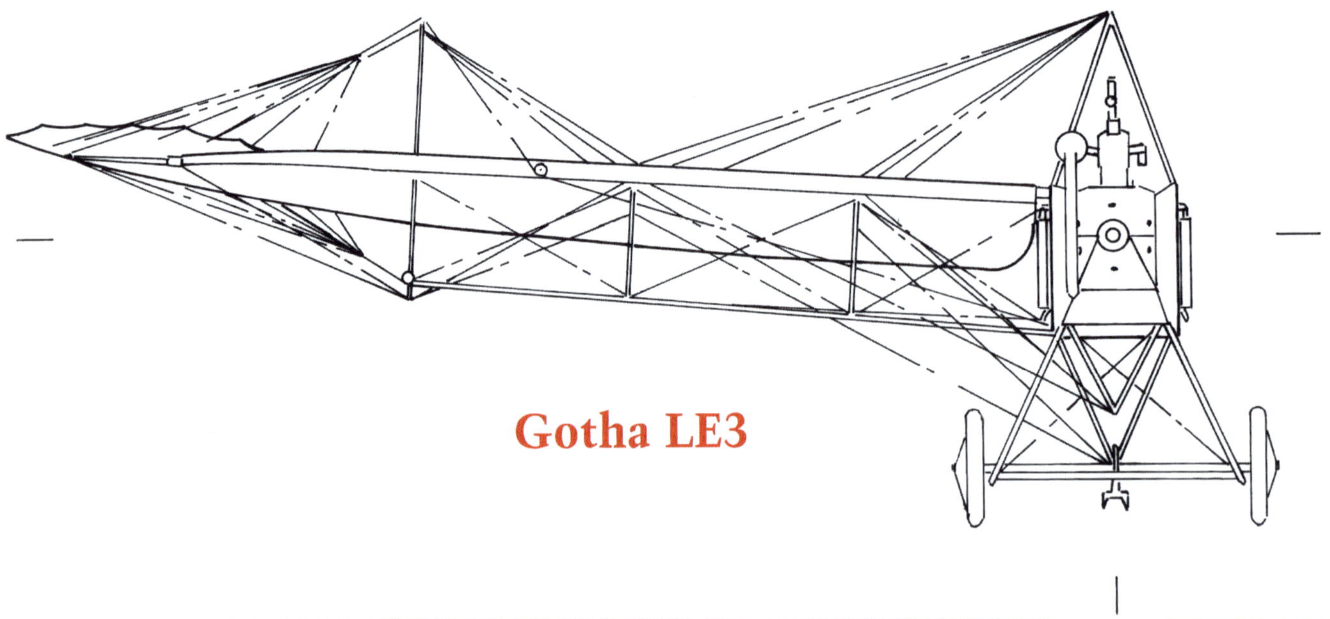

Gotha LE3

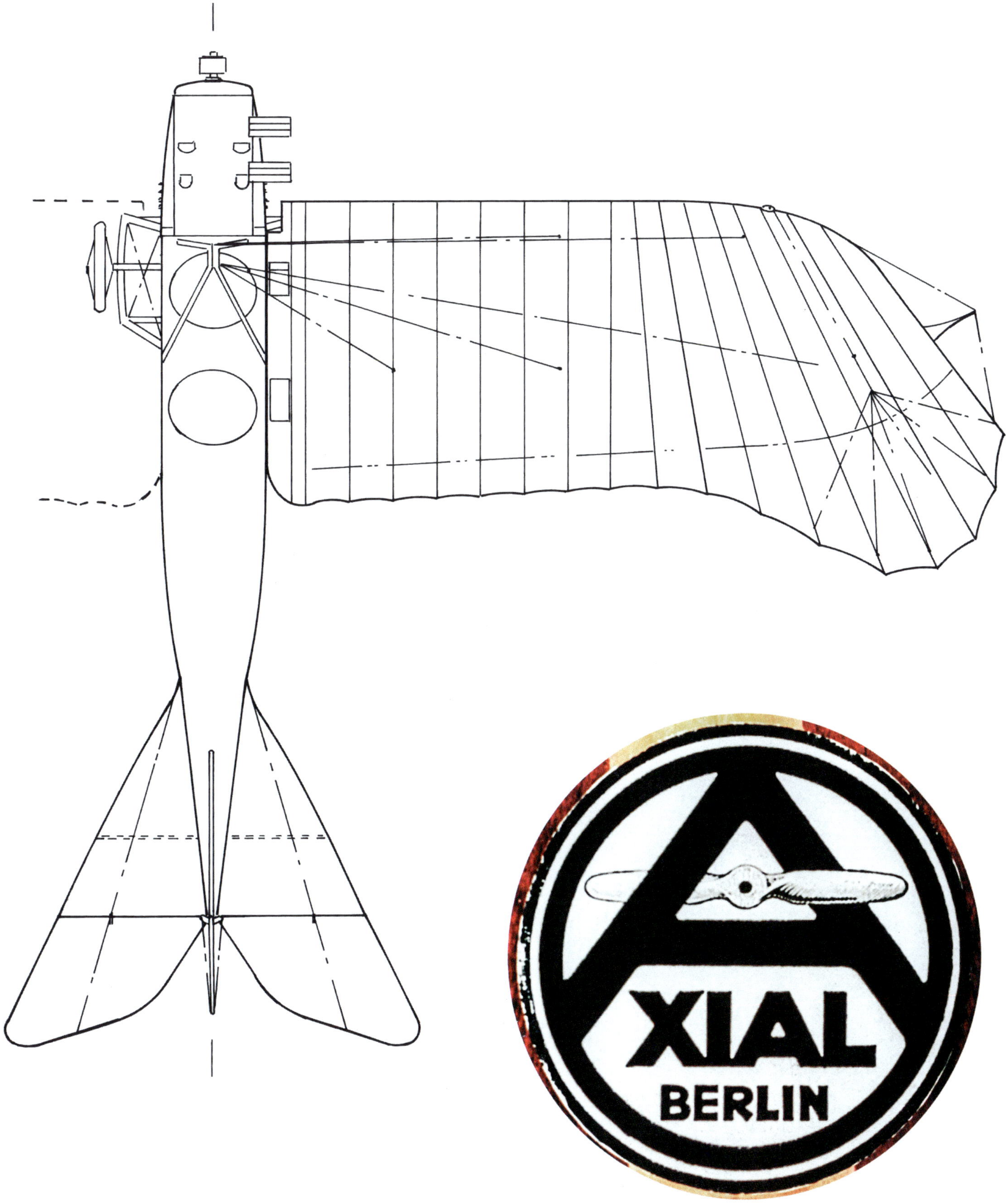

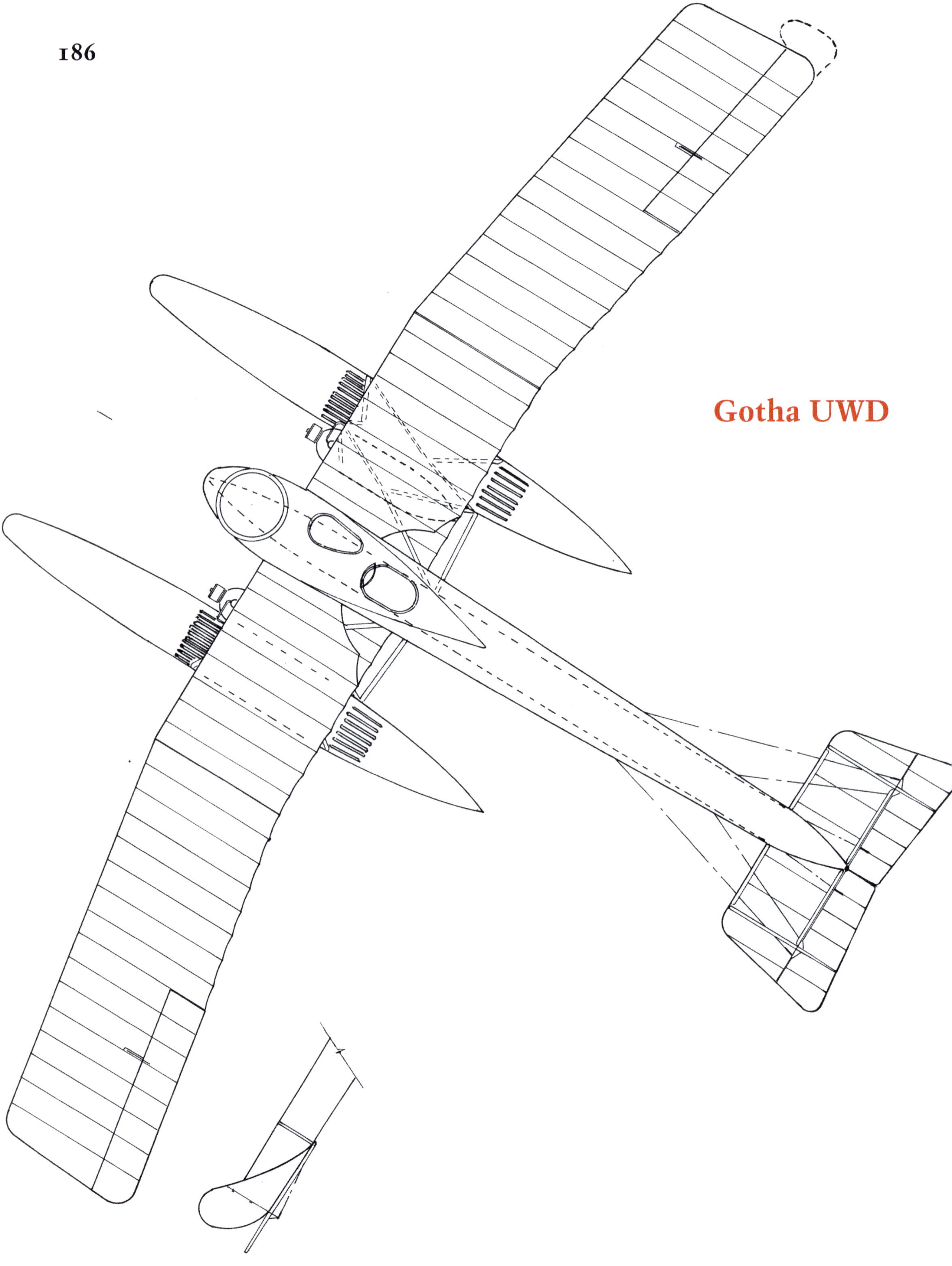

Gotha UWD

Gotha UWD

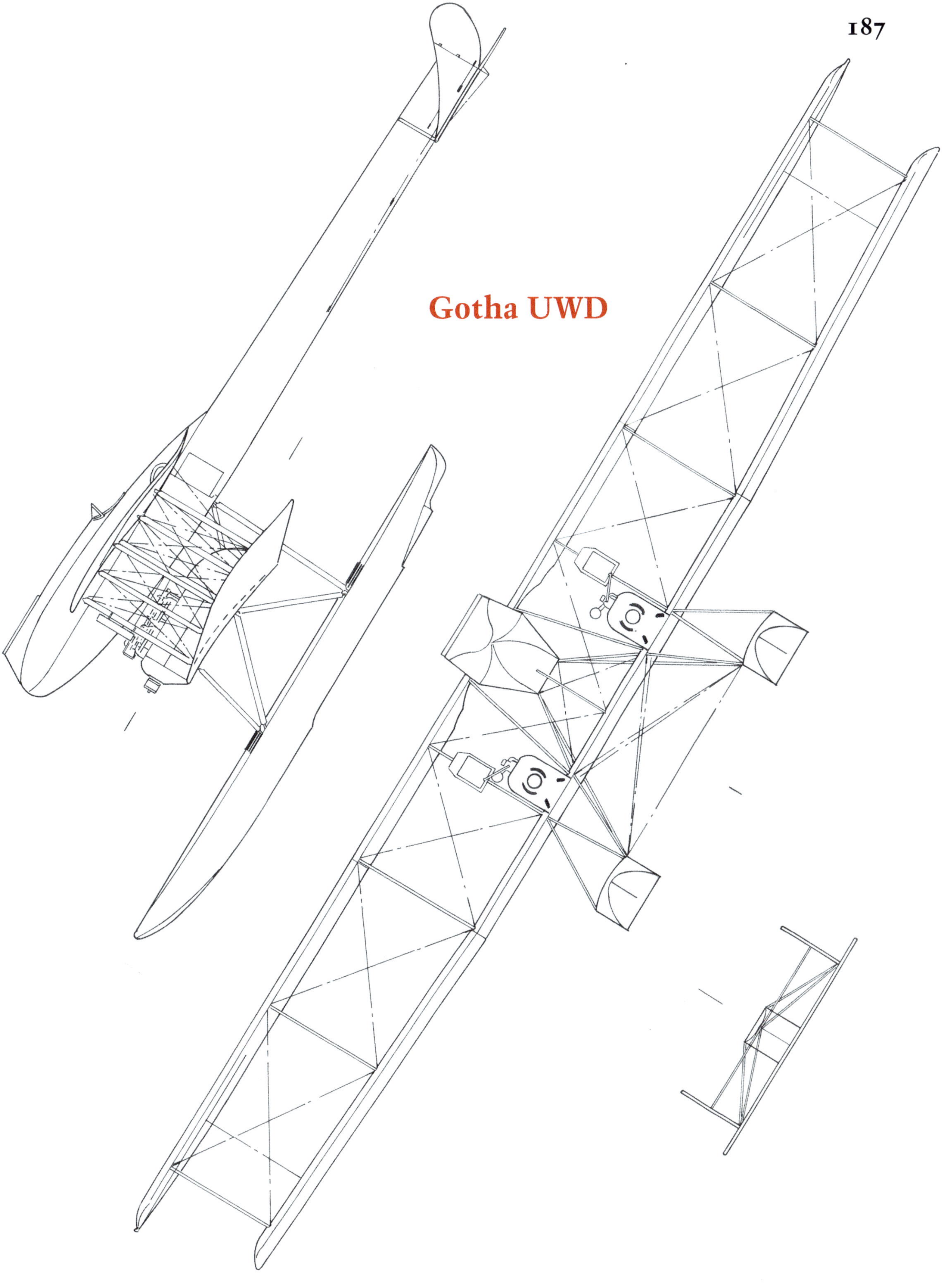

Gotha WD7

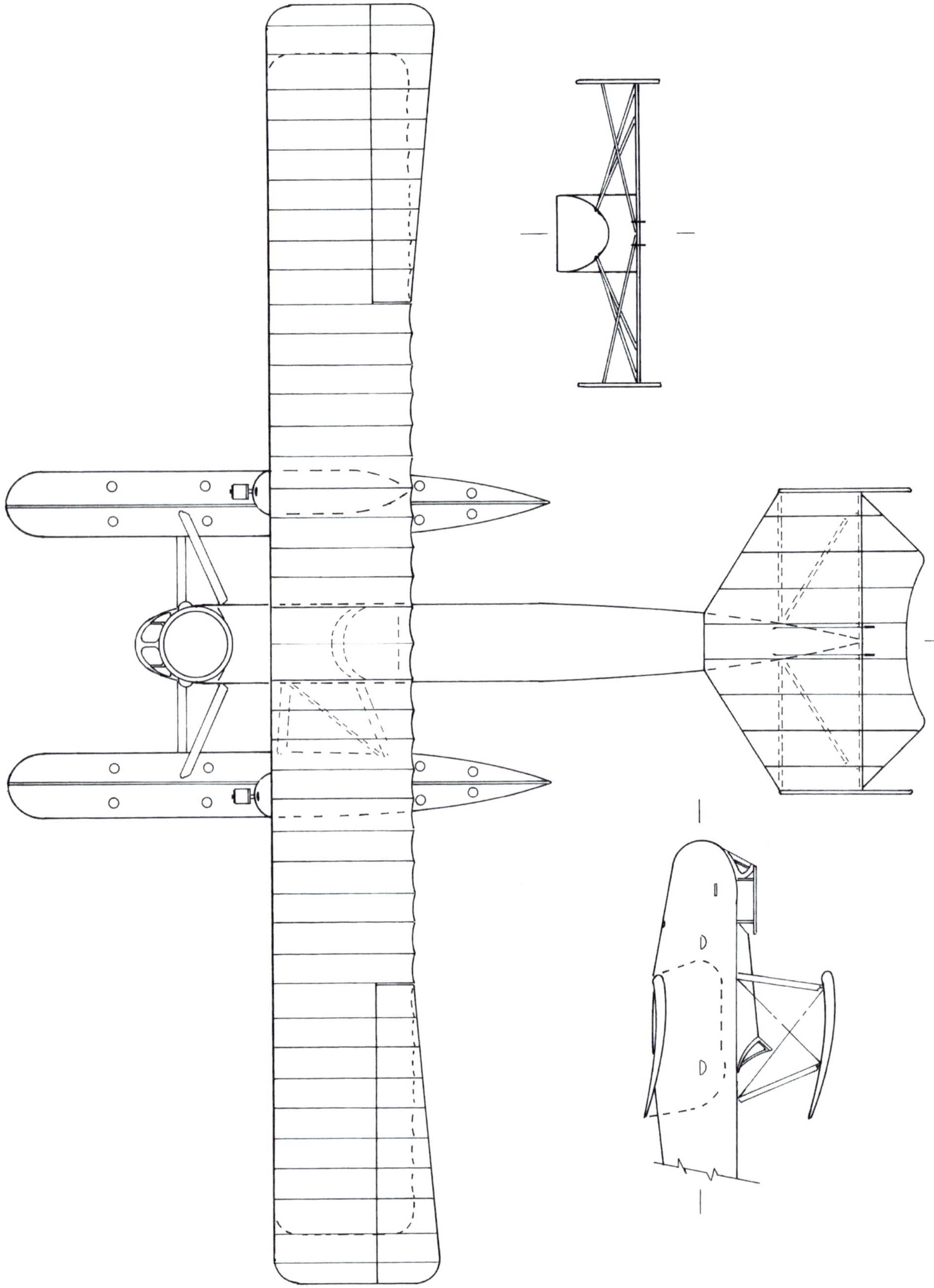

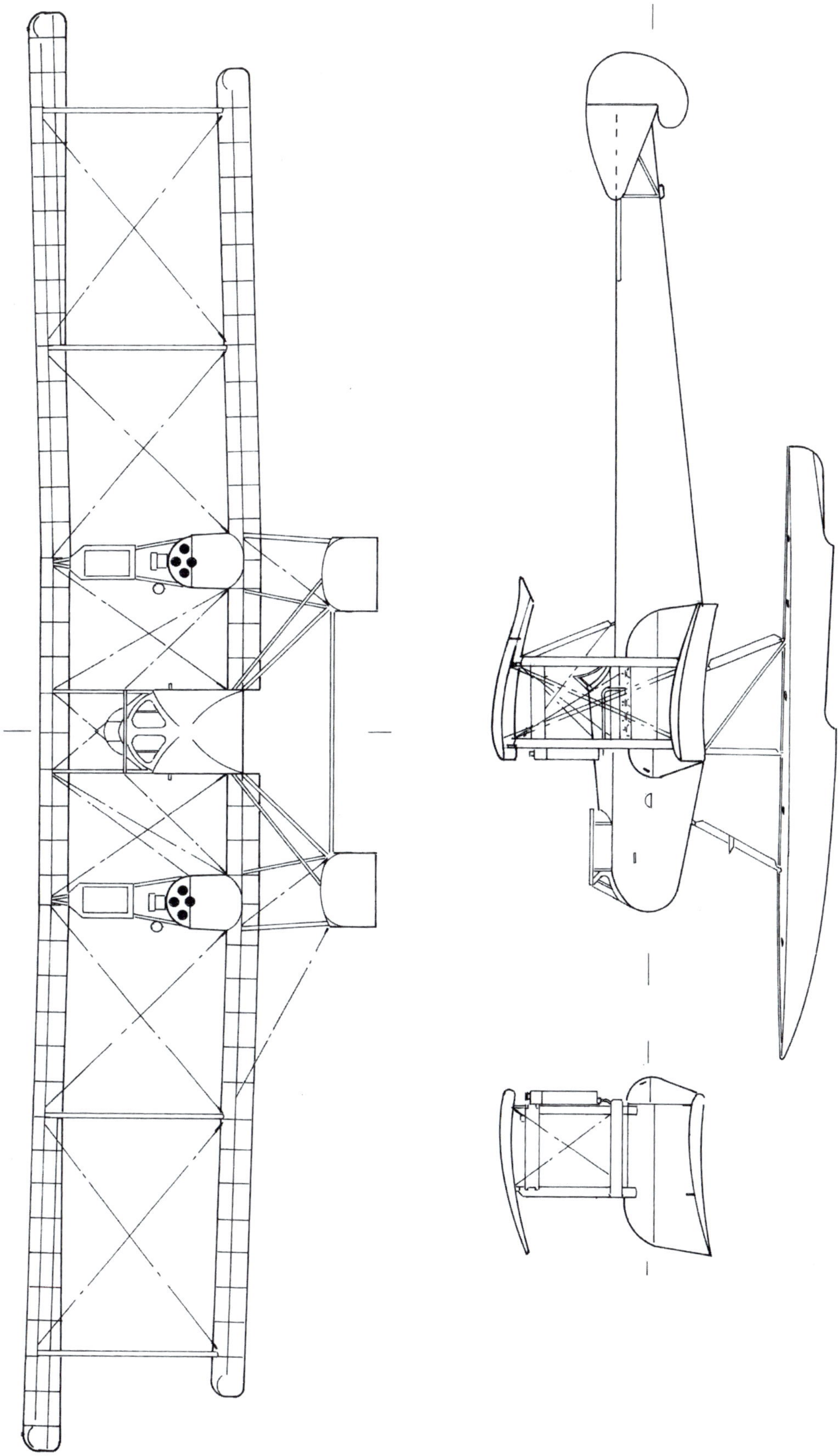

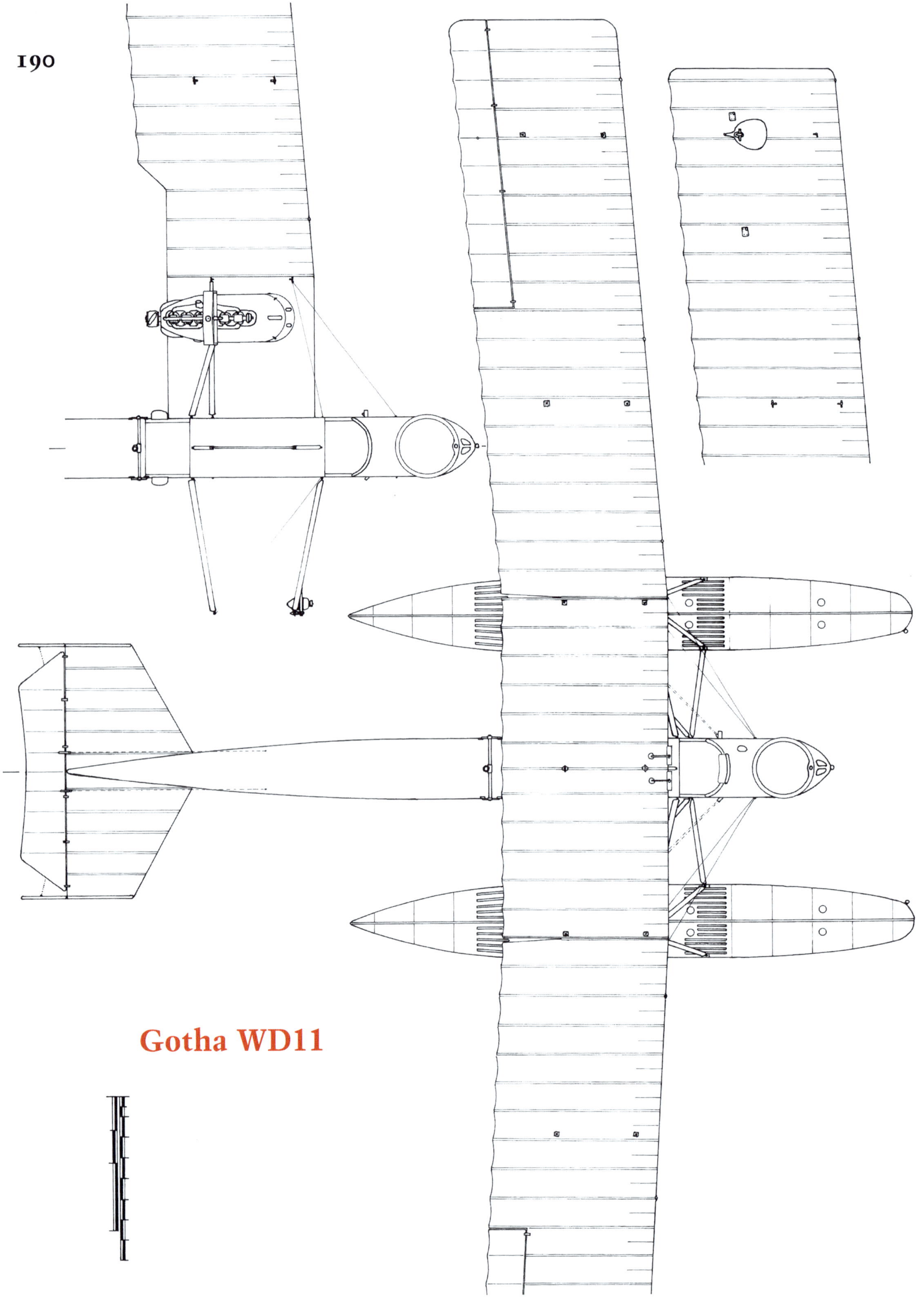

Gotha WD11

Gotha WD11

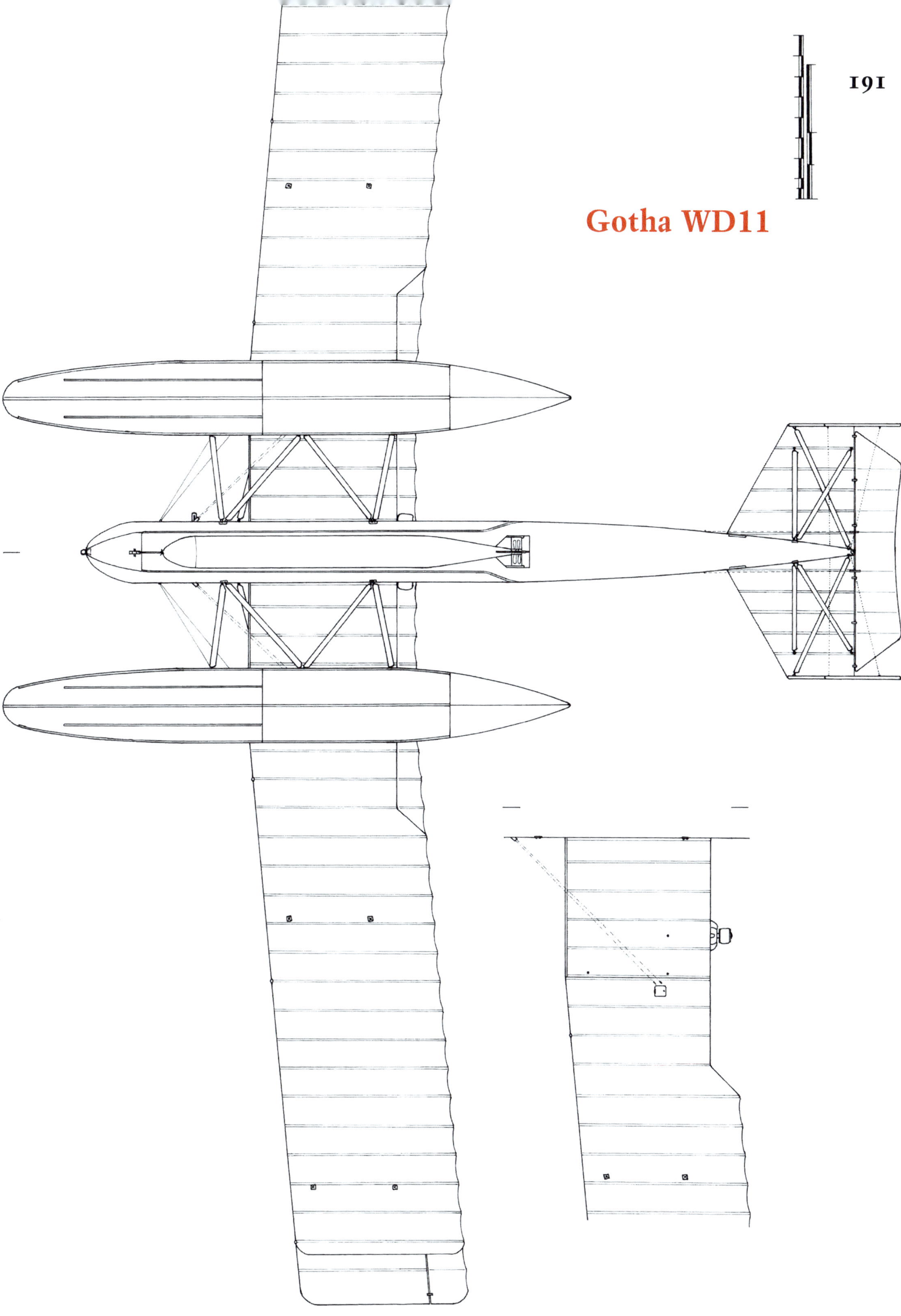

Gotha WD11

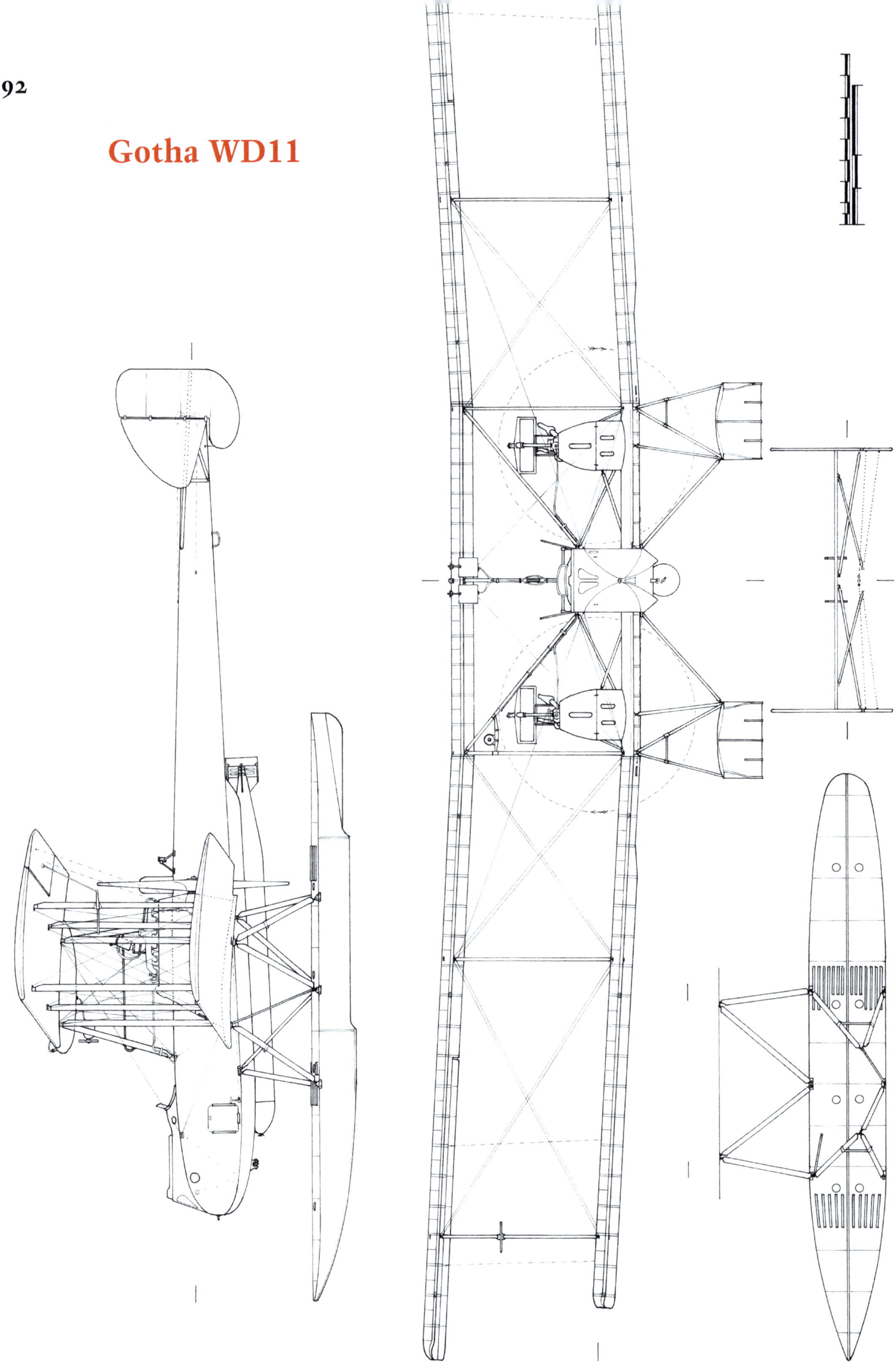

Gotha WD14

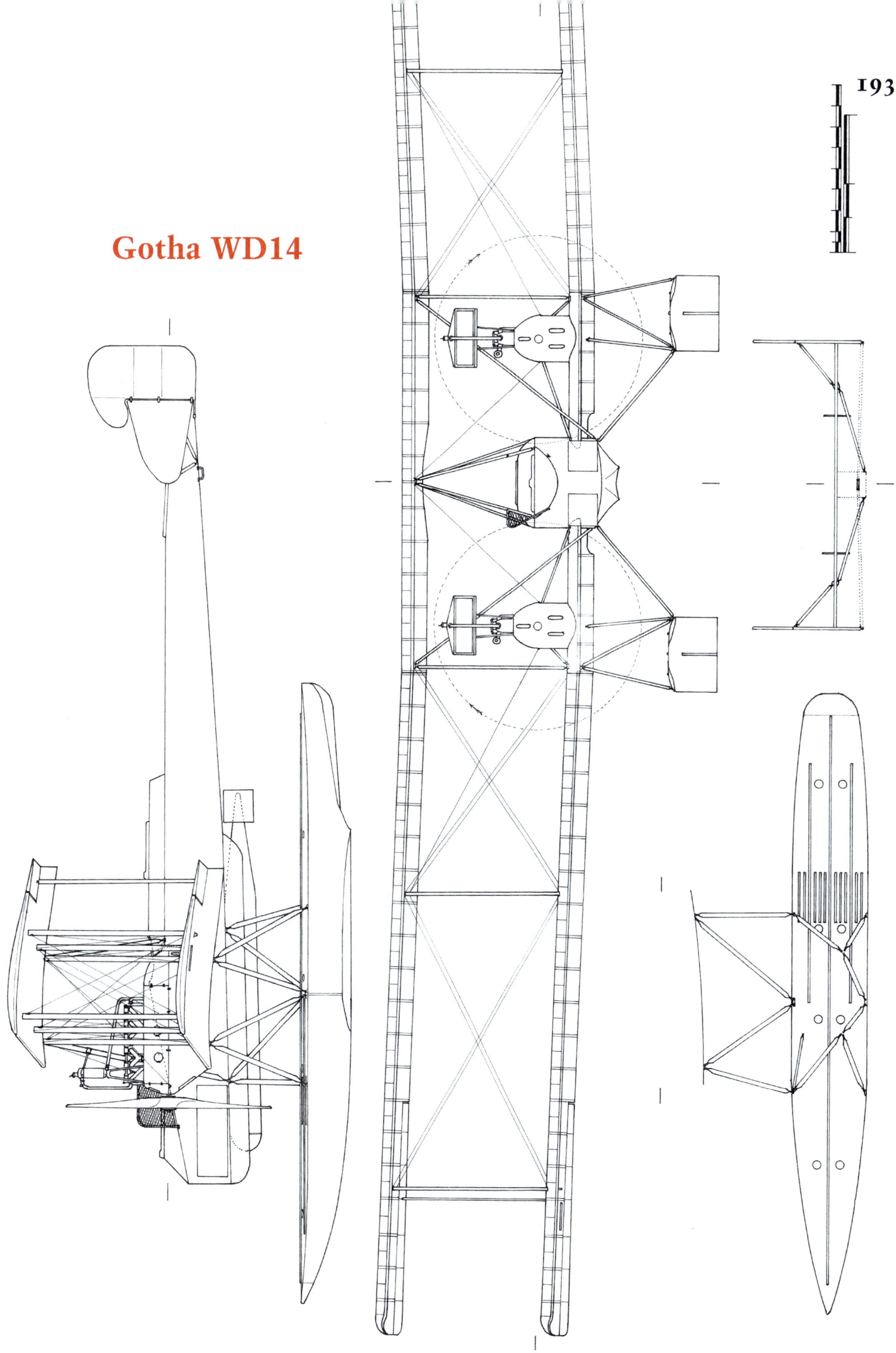

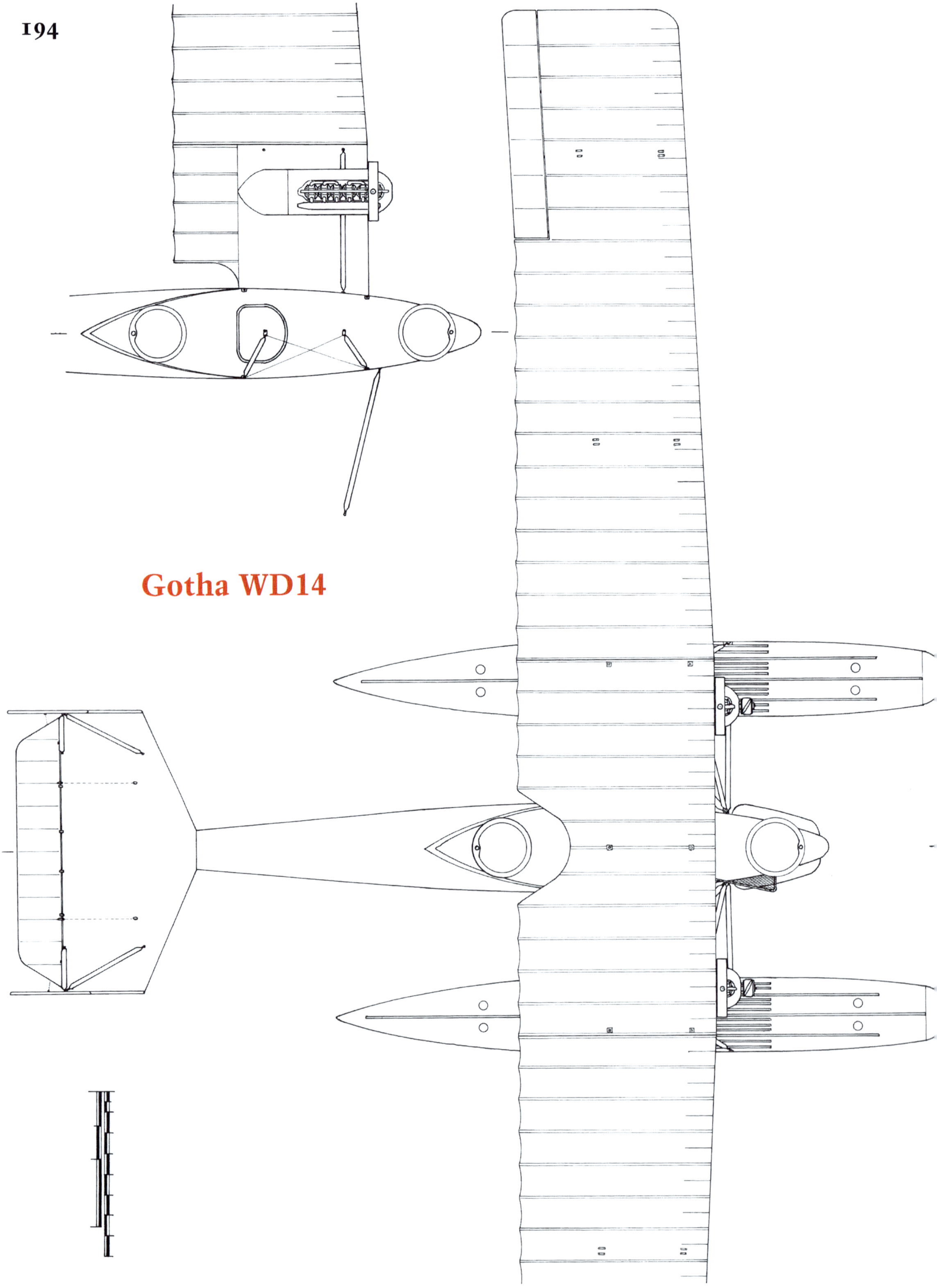

Gotha WD14

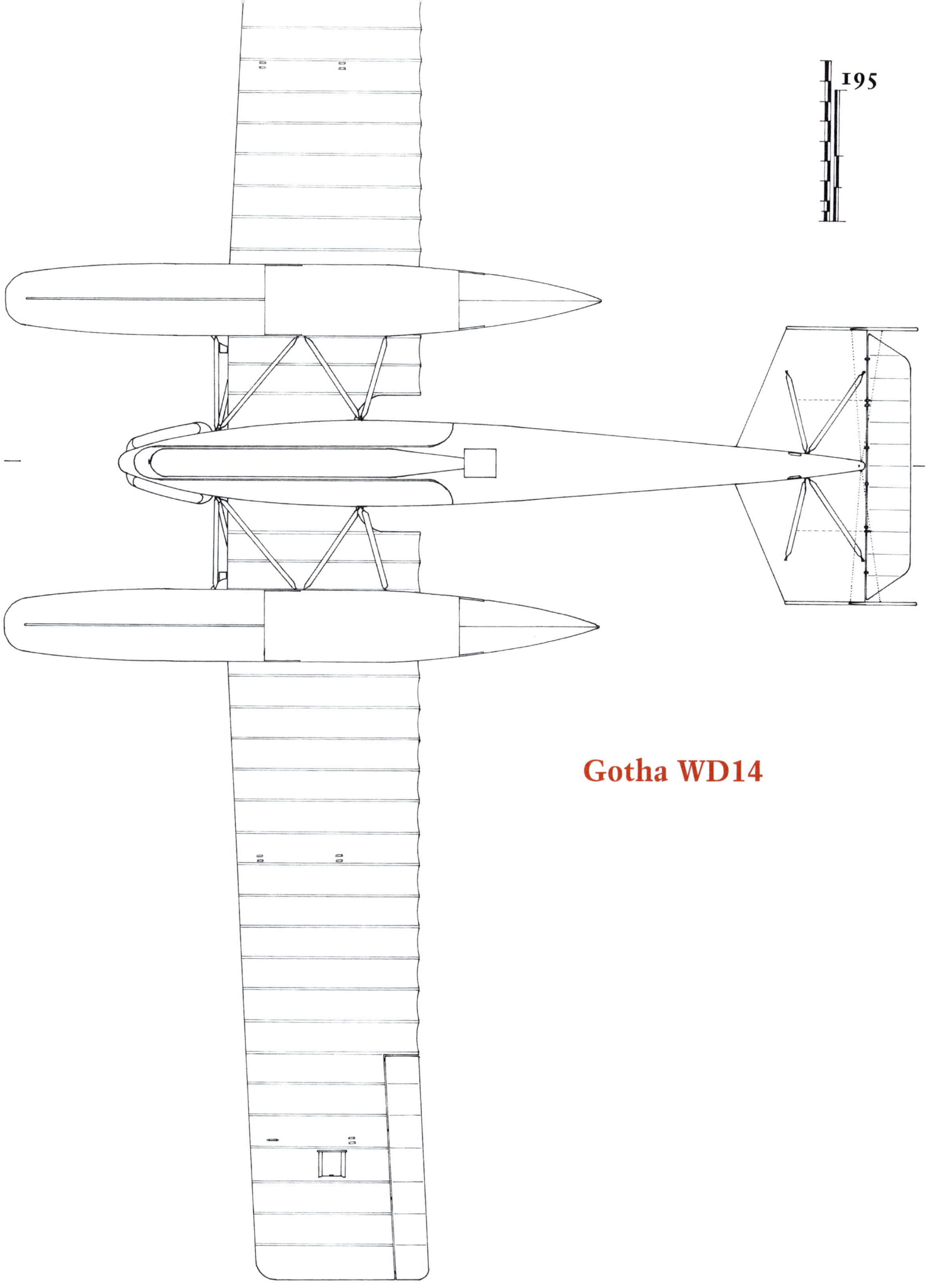

Gotha WD14

Afterword

Without doubt the most notable Gotha crewman was *Hptm*. Ernst Brandenburg, commander of *KG*3, who lead the daylight attacks on London. Brandenburg started the war as an *Oblt*. in the 6th West Prussian Infantry Regiment Nr.149; after being wounded and declared unfit for further duty in the trenches, he became an observer. In January 1917 he was asked to form *KG*3, also known as the *England Geschwader* for its mission of bombing England, especially London. Brandenburg led *KG*3 on the first successful daylight attack on London on 13 June 1917; the next day Brandenburg was promoted to *Hptm*. and awarded the *Pour le Mérite* for this achievement.

The day after receiving the *Pour le Mérite,* Brandenburg was being flown back to *KG*3 in an Albatros two-seater by *Oblt. Freiherr* von Trotha when the engine failed on take-off. Brandenburg was severely injured in the resulting crash and one of his legs had to be amputated. *Hptm*. Kleine assmed command of *KG*3 on 23 June, but after his death in action on 12 December, Brandenburg resumed command of *KG*3, now called *Bogohl* 3, until the armistice.

Above: *Hptm*. Ernst Brandenburg (1883–1952) photographed after receiving the *Pour le Mérite* for leading the first successful Gotha bombing attack on London.

Selected Bibliography

British Vessels Lost at Sea, 1914–1918, HMSO, 1919

Cole, Christopher, Cheesman, E.F., *The Air Defence of Britain 1914–1918*, Putnam, 1984.

Grosz, P.M., Haddow, G., & Schiemer, P., *Austro-Hungarian Army Aircraft of World War One*, Flying Machines Press, 1993.

Grosz, P.M., *Gotha!*, Albatros Productions, Ltd., 1994.

Grosz, P.M., *Gotha G.I*, Albatros Productions, Ltd., 2000.

Grosz, P.M. "Evolution of the Gotha G.VII, G.VIII and G.IX." World War I Aero No.105, p.20.

Grosz, P.M., *The Taube at War*, Albatros Productions, Ltd., 2004.

Jones, H.A., *The War in the Air*, Vol.IV, 1934, IWM ed. 1998.

Merrill, Glen, *Gross- und Riesenflugzeug Markierungen, 1916–18 (1) Gotha*, Americal/Gryphon No. 51 & 52

Owers, Colin A., *Late Gotha Bombers*, Albatros Productions, Ltd., 2010.

Gray, P., & Thetford, O., *German Aircraft of the First World War*, Doubleday & Company, Inc., Second Revised Edition, 1970.

Flugzeug der GFW (Gothaer Waggonfabrik AG), Luftfahrt-Verlag Walter Zuerl

www.tayyareci.com/digerucaklar/turkiye/ww1/

Above: The Gotha LD6a with larger fin and rudder. Only one LD6a was built, and it was assigned to the Herzog Carl Eduard Fliegerschule, where it received school number 70. The engine was a 150 hp Benz Bz.III.

Selected Index

Above: Side view of the Gotha LE4; lack of national insignia indicates this is the original configuration before repair. The engine was a 100 hp Mercedes D.I inline six-cylinder.

Above: The Gotha LE4 photographed after repair and modification. During repair the LE4 was modified to have the front and rear lift wires attached to separate pylons in front and behind the undercarriage. After repair and modification, it was assigned to the Herzog Carl Eduard Fliegerschule in Gotha. The nose radiator and horizontal tail with conventional, hinged elevators made the LE4 much more modern in appearance than earlier *Tauben*. However, by the time the LE4 was built in 1914, the *Taube* configuration was obsolete and only a single LE4 was built.

26508509R00110

Made in the USA
Lexington, KY
04 October 2013